U.S. Department of Commerce
National Institute of Standards and Technology

Applied Economics Office
Engineering Laboratory
Gaithersburg, Maryland 20899

Prototype Commercial Buildings for Energy and Sustainability Assessment: Whole Building Energy Simulation Design

Joshua D. Kneifel

Sponsored by:

National Institute of Standards and Technology
Engineering Laboratory

and

U.S. Department of Energy
Federal Energy Management Program

September 2011

U.S. DEPARTMENT OF COMMERCE

Rebecca M. Blank, Acting Secretary

NATIONAL INSTITUTE OF STANDARDS AND TECHNOLOGY

Patrick D. Gallagher, Director

Abstract

Energy efficiency requirements in current commercial building energy codes vary across states, and most states have not yet adopted the newest energy standards. Energy standards currently adopted by states range across all editions of *ASHRAE 90.1* (*-1999, -2001, -2004,* and *-2007*). Some states do not have a code requirement for energy efficiency, leaving it up to the locality or jurisdiction to set their own requirements.

The Applied Economics Office in NIST's Engineering Laboratory has developed an extensive database that allows energy efficiency and sustainability comparisons of alternative building designs based on different editions of *ASHRAE Standard 90.1*. The expansive database is a compilation of multiple data sources, including results from 13 680 whole building energy simulations for 12 commercial building types in 228 cities across all U.S. states, building construction cost databases, energy cost data collected from the Energy Information Administration, and emissions data collected from the Environmental Protection Agency. This report documents the whole building energy simulation designs for the prototype commercial buildings used in this building energy efficiency and sustainability database.

Keywords

Building economics; economic analysis; energy efficiency; commercial buildings

Preface

This report was conducted by the Applied Economics Office in the Engineering Laboratory (EL) at the National Institute of Standards and Technology (NIST). The report is designed to document the assumptions used in creating the whole building energy simulations used for new commercial building prototypes. The intended audience is the building research community, particularly those concerned with energy efficiency in commercial building designs.

Disclaimer

Certain trade names and company products are mentioned in the text in order to adequately specify the technical procedures used. In no case does such identification imply recommendation or endorsement by the National Institute of Standards and Technology, nor does it imply that the products are necessarily the best available for the purpose.

Disclaimer

The policy of the National Institute of Standards and Technology is to use metric units in all of its published materials. Because this report is intended for the U.S. construction industry that uses U.S. customary units, it is more practical and less confusing to include U.S. customary units as well as metric units. Measurement values in this report are therefore stated in metric units first, followed by the corresponding values in U.S. customary units within parentheses.

Acknowledgements

The author would like to thank the NIST Engineering Laboratory and the Department of Energy Federal Energy Management Program for their support of the project. The author also wishes to thank all those who contributed ideas and suggestions for this report. They include Ms. Barbara Lippiatt and Dr. Robert Chapman of EL's Applied Economics Office, Dr. William Healy of EL's Building Environment Division, and Dr. Nicos S. Martys of EL's Materials and Construction Research Division. A special thanks to Nick Long and the EnergyPlus Team for generating the initial energy simulations. Thanks to Brian Presser for altering the heating and cooling equipment in the initial whole building energy simulations to replicate the RSMeans prototype buildings and meet *ASHRAE 90.1* efficiency requirements, and generating the final simulations used in the database.

Author Information

Joshua D. Kneifel
Economist
National Institute of Standards and Technology
Engineering Laboratory
100 Bureau Drive, Mailstop 8603
Gaithersburg, MD 20899-8603
Tel.: 301-975-6857
Email: joshua.kneifel@nist.gov

Table of Contents

List of Figures

List of Tables

List of Acronyms

Acronym	Definition
ACH	Air Changes Per Hour
AEO	Applied Economics Office
ASHRAE	American Society of Heating, Refrigerating and Air-Conditioning Engineers
CBECS	Commercial Building Energy Consumption Survey
DOE	Department of Energy
EEFG	EnergyPlus Example File Generator
EL	Engineering Laboaratory
FEMP	Federal Energy Management Program
FERC	Federal Energy Regulatory Commission
HVAC	Heating, Ventilating, and Air Conditioning
IECC	International Energy Code Council
LEC	Low Energy Case
NIST	National Institute of Standards and Technology
PNNL	Pacific Northwest National Laboratory
SEER	Seasonal Energy Efficiency Ratio
SHGC	Solar Heat Gain Coefficient

1 Introduction

1.1 Background

Energy efficiency requirements in current commercial building energy codes vary across states, and most states have not yet adopted the newest energy standards. Energy standards currently adopted by states range across all versions of *ASHRAE 90.1* (*-1999, -2001, -2004,* and *-2007*). Some states do not have a code requirement for energy efficiency, leaving it up to the locality or jurisdiction to set their own requirements. There may be significant energy and cost savings to be realized by states if they were to adopt more energy efficient commercial building energy standards.

The Applied Economics Office in NIST's Engineering Laboratory has developed an extensive database that allows energy efficiency and sustainability comparisons of alternative building designs based on different editions of *ASHRAE Standard 90.1*. The expansive database is a compilation of multiple data sources, including results from 13 680 whole building energy simulations for 12 commercial building types in 228 cities across all U.S. states, building construction cost databases, energy cost data collected from the Energy Information Administration, and emissions data collected from the Environmental Protection Agency.

1.2 Purpose

The purpose of this report is to document the whole building energy simulations used to estimate the energy use of the new commercial building prototype designs implemented in the building energy efficiency and sustainability database. Future analyses of the database will be able to cite these detailed energy simulation assumptions and summarize the documentation, allowing the research to focus on the results.

1.3 Literature Review

The National Renewable Energy Laboratory (NREL) has documented 16 commercial prototypical buildings for different climate zones.[1] These 256 reference commercial buildings are based on the Commercial Building Energy Consumption Survey (CBECS) database and the *American Society of Heating, Refrigerating and Air-Conditioning Engineers (ASHRAE) 90.1 Standard*, and represent about 70 % of the U.S. building stock of conditioned floor area. These prototypes have become the basis for significant research in the analysis of energy efficiency measures in commercial buildings. In a similar manner, this report will define prototypical buildings that can be used to research energy efficiency and sustainability in commercial buildings.

---

[1] Field, Deru, and Studer (2010)

1

1.4   Approach

This report documents the assumptions used to create the whole building energy simulations that are used to estimate the energy use for the new commercial building prototype designs, which include building geometry, building construction, occupancy type, lighting, electrical equipment, heating, ventilation, and air conditioning (HVAC) equipment, infiltration, and ventilation. The initial whole building energy simulations were created using the *EnergyPlus Example File Generator (EEFG)*. These initial simulations were altered to match the HVAC equipment types in the prototype buildings from the cost database.

2 Commercial Building Designs

Current state energy codes are based on different iterations of the *International Energy Conservation Code (IECC)* or *ASHRAE 90.1* Standards, which have requirements that vary based on a building's characteristics and the climate zone of the location. For this study, the *ASHRAE* Standard equivalent design is used to meet current state energy codes and to define the alternative building designs.

Table 2-1 shows that current commercial building energy codes vary by state. In a few instances, local jurisdictions have adopted energy standards that are more stringent than the state energy codes.[2] These cities are also included in Table 2-1.

Table 2-1 Energy Code by State and City Exceptions

Location	Energy Code	Location	Energy Code	Location	Energy Code
AK	None	IN	2007	NV	2004
AL	None	KS	None	NY	2007
Huntsville	2001	KY	2007	OH	2004
AR	2001	LA	2007	OK	None
AZ	None	MA	2007	OR	2007
Flagstaff	2004	MD	2007	PA	2007
Phoenix	2004	ME	2007	RI	2007
Tucson	2004	MI	2007	SC	2004
CA	2007	MN	2004	SD	None
CO	2001	MO	None	Huron	2001
Grand Junction	2004	St Louis	2001	TN	2004
CT	2004	MS	None	TX	2007
DE	2007	MT	2007	UT	2007
FL	2007	NC	2004	VA	2007
GA	2007	ND	None	VT	2004
HI	2004	NE	2007	WA	2007
IA	2007	NH	2007	WI	2007
ID	2007	NJ	2007	WV	2001
IL	2007	NM	2007	WY	None

Note: Some city ordinances require energy codes that exceed state energy codes.

State energy codes vary from no statewide code to *ASHRAE 90.1-2007* with some regional trends shown in Figure 2-1. The states in the central U.S. tend to wait longer to adopt newer *ASHRAE 90.1* Standards. However, there are many cases in which energy codes of neighboring states vary drastically. For example, Illinois, Iowa, and Nebraska have adopted *ASHRAE 90.1-2007* while Kansas, Missouri, and Oklahoma do not have

---

[2] Local and jurisdictional requirements are obtained from the Database of State Incentives for Renewables and Efficiency (DSIRE). State energy code requirements targeting only public buildings and green standards are ignored in this study.

any energy code, and Arkansas has adopted *ASHRAE 90.1-2001*, and Tennessee have adopted *ASHRAE 90.1-2004*.

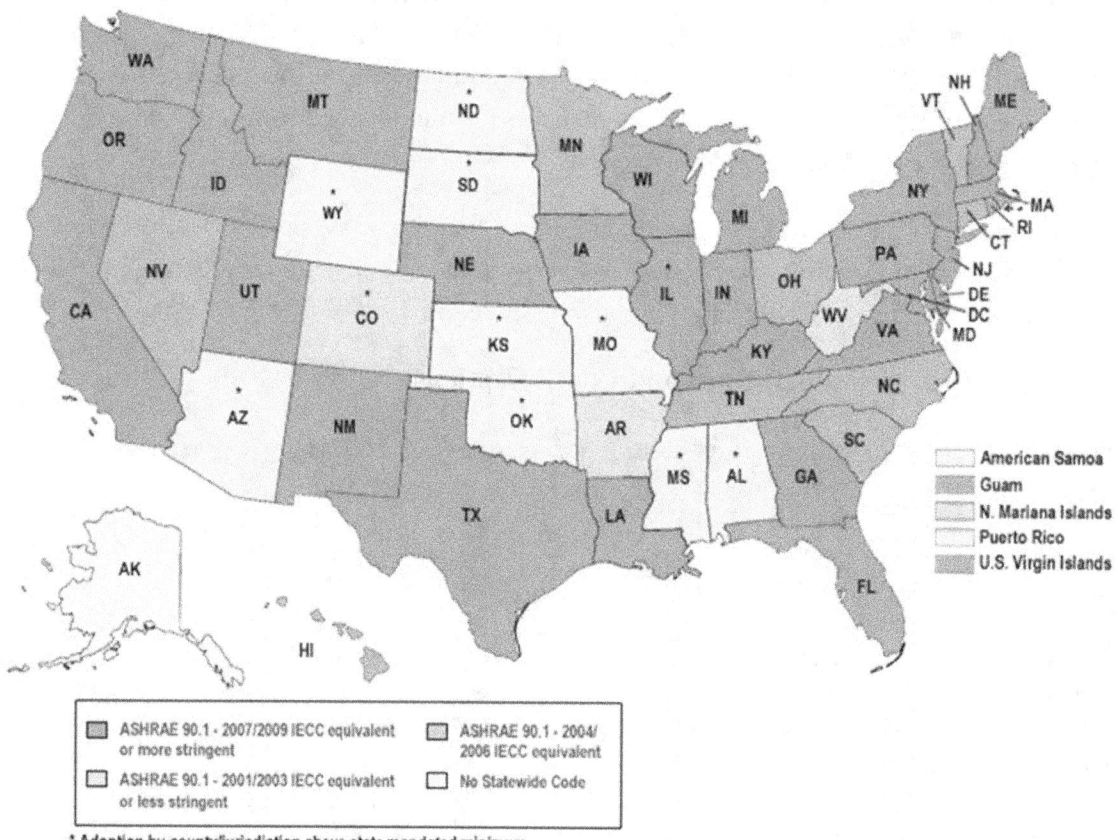

American Samoa
Guam
N. Mariana Islands
Puerto Rico
U.S. Virgin Islands

ASHRAE 90.1 - 2007/2009 IECC equivalent or more stringent

ASHRAE 90.1 - 2004/ 2006 IECC equivalent

ASHRAE 90.1 - 2001/2003 IECC equivalent or less stringent

No Statewide Code

* Adoption by county/jurisdiction above state mandated minimum

Figure 2-1 State Commercial Energy Codes[3]

The energy simulations are designed to meet the requirements for each of the editions of *ASHRAE 90.1* (*-1999, -2001, -2004,* and *-2007*) and an additional building design option defined as the "Low Energy Case" (LEC), which goes beyond *ASHRAE 90.1-2007* in a number of ways. The LEC design increases the thermal efficiency of insulation and windows beyond *ASHRAE 90.1-2007*, reduces the lighting power density, and adds daylighting and window overhangs. The LEC design assumes the same HVAC equipment efficiency as required by *ASHRAE 90.1-2007*. For this study, ASHRAE 90.1-1999 is assumed to be "common practice," and is used for the building design requirements in states with no statewide energy code.

The ASHRAE 90.1 requirements vary depending where the building is located. For this reason energy simulations are designed for 228 cities across the United States across all

---

[3] Obtained from the DOE Building Technologies Program September 20, 2011 (http://www.energycodes.gov/states/maps/commercialStatus.stm).

4

states and climate zones. The cities, shown as dots within a circle, and color-coded ASHRAE climate zones are shown in Figure 2-2. These cities are selected for three reasons. First, the cities are spread out to represent the entire United States, and represent as many climate zones in each state as possible. Second, the locations cover all the major population centers in the country. Third, multiple locations for a climate zone within a state are included to allow building costs to vary for each building design. See Table B-1 in Appendix B for the entire list of cities and their climate zones.

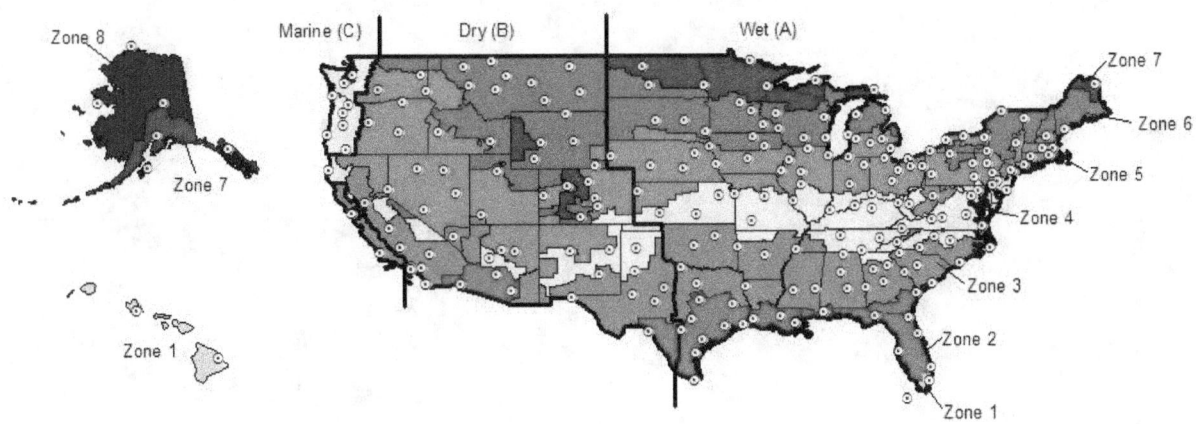

Figure 2-2 Cities and ASHRAE Climate Zones

5

3 Energy Simulation Design

The EnergyPlus whole building energy simulations are generated by the *EnergyPlus Example File Generator (EEFG)*. The *EEFG* input form is displayed in Figure A-1 in Appendix A. The *EEFG* narrows down a building's description into simple, high-level characteristics: building geometry, orientation, number of floors, floor height, building type, wall and roof construction type, window-to-wall ratio, and location. The remaining building parameters are defined based on the chosen building energy standard and some default values based on the building type.

The original simulations obtained from the *EEFG* assumed a unitary HVAC system with gas heat for all building types. The HVAC system in the simulation was replaced with a system that best represents the system defined in the commercial prototype buildings in the cost database, the RSMeans CostWorks Square Foot Cost Estimator (SFCE). Figure A-2 and Figure A-3 in the Appendix show the RSMeans CostWorks SFCE input and output forms for the construction costs of a commercial building prototype. The cooling and heating systems defined in the RSMeans CostWorks SFCE for each building type are described in Chapter 4.

The efficiency of a particular piece of HVAC equipment is determined by the selected building design, which in this report equates to the selected edition of *ASHRAE 90.1*. The building envelope design determines the capacity necessary for the HVAC equipment to meet the thermal load, which in turn requires the simulation of the design days. Based on the HVAC equipment capacity and the HVAC equipment type, the *ASHRAE 90.1* efficiency requirement can be determined. The HVAC equipment types and efficiency requirements are discussed in Section 4.10.

The simulation assumes parameter values for the exterior envelope that represent the performance of each surface as a single material. For example, a window is represented as a single layer with parameter values that represent the combined performance characteristics of each layer of the window. The individual components of the window (e.g. panes, coatings, films, gas fill, etc.) are not specified in the simulation, only the overall U-factor, Solar Heat Gain Coefficient (SHGC), and Visual Transmittance (VT) of the window.

4 Energy Simulation Assumptions

The building characteristics in Table 4-1 describe the 12 building types used in this study, which include 2 dormitories, 2 apartment buildings, 1 hotel, 3 office buildings, 2 schools, 1 retail store, and 1 restaurant, and represent 46 % of the U.S. commercial building stock floor space.[4] The prototype buildings range in size from 465 m^2 (5000 ft^2) to 41 806 m^2 (450 000 ft^2). The building abbreviations defined in Table 4-1 are used to represent the building types in tables throughout this report.

Table 4-1 Building Characteristics

Building Type	Bldg. Abbr.	Floors	Floor Height m (ft)	Wall	Roof†	Pct. Glazing	Floor Area m^2 (ft^2)	CBECS Occupancy Type	U.S. Floor Space (%)
Dormitory	DORMI04	4	3.66 (12)	Mass*	IEAD	20 %	3097 (33 333)	Lodging	7.1 %
Dormitory	DORMI06	6	3.66 (12)	Steel	IEAD	20 %	7897 (85 000)		
Hotel	HOTEL15	15	3.05 (10)	Steel	IEAD	100 %	41 806 (450 000)		
Apartment	APART04	4	3.05 (10)	Mass*	IEAD	12 %	2787 (30 000)		
Apartment	APART06	6	3.15 (10)	Mass*	IEAD	14 %	5574 (60 000)		
School, Elem.	ELEMS01	1	4.57 (15)	Mass*	IEAD	25 %	4181 (45 000)	Education	13.8 %
School, High	HIGHS02	2	4.57 (15)	Mass*	IEAD	25 %	12 077 (130 000)		
Office	OFFIC03	3	3.66 (12)	Mass	IEAD	20 %	1858 (20 000)	Office	17.0 %
Office	OFFIC08	8	3.66 (12)	Mass*	IEAD	20 %	7432 (80 000)		
Office	OFFIC16	16	3.05 (10)	Steel	IEAD	100 %	24 155 (260 000)		
Retail Store	RETAIL1	1	4.27 (14)	Mass*	IEAD	10 %	743 (8000)	Mercantile**	6.0 %
Restaurant	RSTRNT1	1	3.66 (12)	Wood	IEAD	30 %	465 (5000)	Food Service	2.3 %

†IEAD = Insulation Entirely Above Deck
*Mass walls include all masonry wall construction types
**Only includes non-mall floor area.

The occupancy types are based on the 1999 Commercial Building Energy Consumption Survey (CBECS) "principal building activity" categories.[5] The principal building activity determines the maximum people density, electrical plug density, and lighting power density, and the correlating density schedules. The *EEFG* "Smart Default" values are used for all densities in the simulations. Table 4-2 summarizes each building type's maximum occupancy, lighting, and equipment density. The density schedules, based on the percentage of peak density, are defined later in this section for each building type.

---

[4] Based on the Commercial Building Energy Consumption Survey (CBECS) database. The NREL prototypes include 4 additions building types, and represent 24 % more (70 %) of the U.S. building stock.

[5] CBECS categories are listed in

Table B-2.

Maximum number of occupants varies between 1 person per 7.0 m^2 (75 ft^2) to 1 person per 27.9 m^2 (300 ft^2). The simulations assume that an occupant's "activity level" sets a person's internal heat gains at 120 W/person. These heat gains from occupants are broken up into 30 % radiant and 70 % sensible loads.

The maximum interior lighting density and daily schedule varies based on the occupancy type and the edition of *ASHRAE 90.1* or LEC building designs. All building types are assumed to have no external lighting loads.

Table 4-2 Occupancy, Lighting, and Equipment

Bldg. Abbr.	ASHRAE 90.1 Occupancy Type	Max. Occ.	m^2 (ft^2) Per Occupant	Lighting W/m^2 (W/ft^2)	Equipment W/m^2 (W/ft^2)
DORMI04	Dormitory	132	23.2 (250)	8.6-18.3 (0.8-1.7)	2.69 (0.25)
DORMI06	Dormitory	342	23.2 (250)	8.6-18.3 (0.8-1.7)	2.69 (0.25)
HOTEL15	Hotel	1800	23.2 (250)	8.6-18.3 (0.8-1.7)	2.69 (0.25)
APART04	Dormitory	120	23.2 (250)	8.6-18.3 (0.8-1.7)	2.69 (0.25)
APART06	Dormitory	240	23.2 (250)	8.6-18.3 (0.8-1.7)	2.69 (0.25)
ELEMS01	School	602	7.0 (75)	10.8-16.1 (1.0-1.5)	5.38 (0.50)
HIGHS02	School	1740	7.0 (75)	10.8-16.1 (1.0-1.5)	5.38 (0.50)
OFFIC03	Office	72	25.5 (275)	8.6-14.0 (0.8-1.3)	8.07 (0.75)
OFFIC08	Office	288	25.5 (275)	8.6-14.0 (0.8-1.3)	8.07 (0.75)
OFFIC16	Office	944	25.5 (275)	8.6-14.0 (0.8-1.3)	8.07 (0.75)
RETAIL1	Retail	27	27.9 (300)	16.1-20.5 (1.5-1.9)	2.69 (0.25)
RSTRNT1	Dining: Fast Food	50	9.3 (100)	14.0-19.4 (1.3-1.8)	1.08 (0.10)

The maximum electrical equipment load varies between 1.08 W/m^2 (0.10 W/ft^2) to 8.07 W/m^2 (0.75 W/ft^2). The heat gain fractions from equipment are the same across all building types: 50 % of the energy is converted to radiant heat, 50 % is converted to sensible heat, and 0 % of the energy is converted to latent heat or lost to the outside environment.

The heating and cooling setpoint temperatures and setpoint schedules vary by occupant type. These setpoints are described in detail in Section 4.1 through Section 4.12.

The HVAC systems are automatically sized for each location by EnergyPlus based on three design day outdoor conditions that are more restrictive than those recommended in the ASHRAE Fundamentals Handbook. The cooling load is based on two sets of design conditions based on the Typical Meteorological Year (TMY2) data: 0.4 % design dry-bulb temperature and mean coincident wet bulb temperature, and 0.4 % design wet-bulb temperature and mean coincident dry bulb temperature. The heating load is based on the 99.6 % dry-bulb design conditions. Both the heating and cooling auto-sizing use a sizing factor of 1.2.

The floor area, number of floors, floor height, wall type, roof type, percent glazing, and HVAC system for each building type are based on the RSMeans CostWorks SFCE default prototype specifications.[6] The cooling system is assumed to run on electricity while the heating system is assumed to run on natural gas. Table 4-3 lists the heating and cooling equipment in each building type.

Table 4-3 HVAC Equipment by Building Type

Building Type	Cooling Equipment	Heating Equipment
DORMI04	Rooftop Packaged Unit	Furnace
DORMI06	Air-Cooled Chiller	Hot Water Boiler
HOTEL15	Water-Cooled Chiller	Hot Water Boiler
APART04	Air-Cooled Chiller	Hot Water Boiler
APART06	Air-Cooled Chiller	Hot Water Boiler
ELEMS01	Split System with Condensing Unit	Hot Water Boiler
HIGHS02	Water-Cooled Chiller	Hot Water Boiler
OFFIC03	Rooftop Packaged Unit	Furnace
OFFIC08	Rooftop Packaged Unit	Furnace
OFFIC16	Water-Cooled Chiller	Hot Water Boiler
RETAIL1	Rooftop Packaged Unit	Furnace
RSTRNT1	Rooftop Packaged Unit	Furnace

Table 4-4 shows how air infiltration and mechanical ventilation rates vary by building type. Infiltration rates range between 0.3 air changes per hour (ACH) to 0.6 ACH. Minimum mechanical ventilation rates range between 0.4 ACH to 1.3 ACH. Total minimum ACH varies between 0.7 ACH to 1.9 ACH.

---

[6] See Figure A-2 and Figure A-3 to see the input and output forms for the RSMeans CostWorks SFCE.

Table 4-4 Air Infiltration and Mechanical Ventilation

Bldg. Abbr.	Building Size m² (ft²)	Floor Height m (ft)	CBECS Occupancy Type	Infiltration (ACH)	Ventilation (ACH)	Total ACH
DORMI04	3097 (33 333)	3.66 (12)	Lodging	0.3	0.4	0.7
DORMI06	7432 (80 000)	3.66 (12)		0.3	0.4	0.7
HOTEL15	41 806 (450 000)	3.05 (10)		0.3	0.5	0.8
APART04	2787 (30 000)	3.05 (10)		0.3	0.5	0.8
APART06	5574 (60 000)	3.15 (10)		0.3	0.5	0.8
ELEMS01	4181 (45 000)	4.57 (15)	Education	0.3	1.0	1.3
HIGHS02	12 077 (130 000)	4.57 (15)		0.3	1.0	1.3
OFFIC03	1858 (20 000)	3.66 (12)	Office	0.3	0.4	0.7
OFFIC08	7432 (80 000)	3.66 (12)		0.3	0.4	0.7
OFFIC16	24 155 (260 000)	3.05 (10)		0.3	0.5	0.8
RETAIL1	743 (8000)	4.27 (14)	Mercantile	0.4	0.6	1.0
RSTRNT1	465 (5000)	3.66 (12)	Food Service	0.6	1.3	1.9

The details of each building type summarized above are given in the remainder of this chapter.

4.1   4-Story Apartment Building

The 4-story apartment building has mass walls, insulation entirely above the roof deck, operable windows, and a window-to-wall ratio of 12 %.[7] The detailed assumptions used in the energy simulations are described below.

4.1.1   Building Envelope

The energy efficiency characteristics of the building envelope are determined by the building's location and the edition of *ASHRAE 90.1*. The window characteristics (U-factor, SHGC, and VT) are based on the *ASHRAE 90.1* requirements for operable windows for 10.1 % to 20.0 % glazing. The wall and roof efficiency characteristics are based on the *ASHRAE 90.1* requirements for residential buildings with above grade, mass wall construction and insulation entirely above the roof deck.

4.1.2   Heating, Ventilation, and Air Conditioning

There are four main aspects to the heating, ventilation, and air conditioning of a building: equipment, operating conditions, air infiltration, and mechanical ventilation. The HVAC equipment is a packaged electric air-cooled chiller and natural gas-fired hot water boiler. Each building type that falls into the "Lodging" CBECS category has the same constant

---

[7] Window to wall ratio is defined as the percentage of the exterior wall area represented by windows.

heating and cooling setpoint temperatures shown in Figure 4-1: 21°C (70°F) for heating and 24°C (76°F) for cooling.

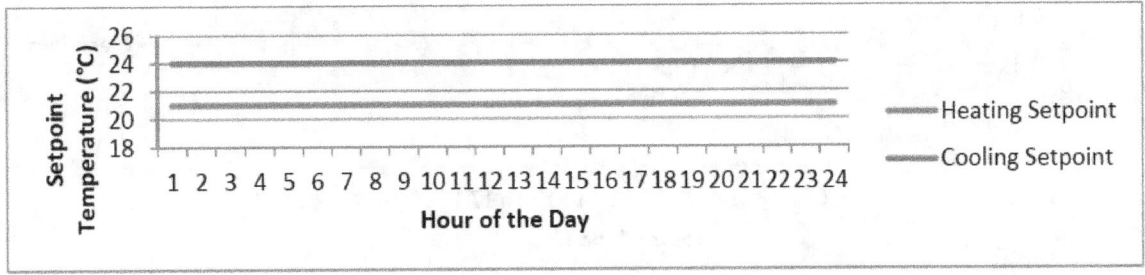

Figure 4-1 APART04 Setpoint Temperature Schedules

Air infiltration and mechanical ventilation are assumed to be constant across all editions of *ASHRAE 90.1*, with an infiltration rate of 0.177 m³/s per floor (0.30 ACH) and minimum mechanical ventilation of 0.284 m³/s per floor (0.48 ACH).

4.1.3   Occupancy, Lighting, and Electrical Loads

The peak occupancy for the 4-story apartment building is assumed to be 120 people or 1 person per 23.2 m² (250 ft²). The schedule in Figure 4-2 shows that the greatest occupancy occurs (i.e., highest fraction of peak) over the nighttime while the lowest occupancy is during the middle of the day.

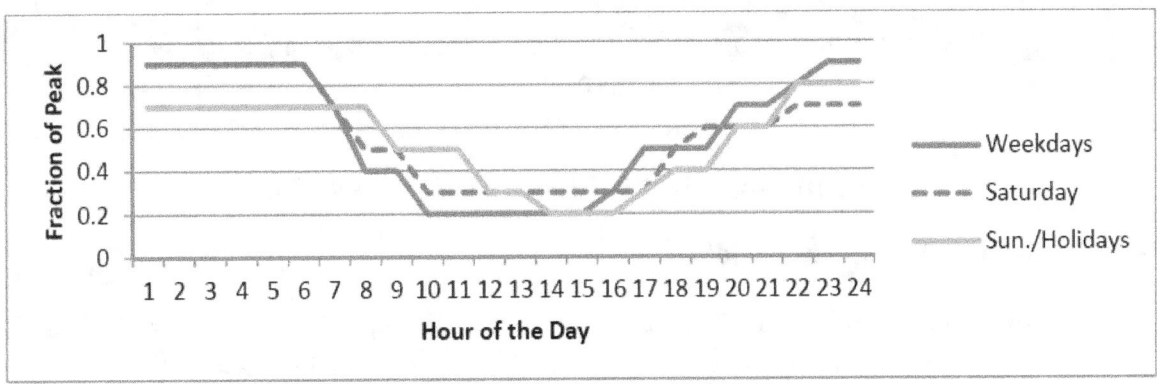

Figure 4-2 APART04 Occupancy Schedule

The energy simulation assumes between 8.6 W/m² (0.8 W/ft²) and 18.3 W/m² (1.7 W/ft²) of lighting density depending on the building design (e.g. edition of *ASHRAE 90.1* or LEC). The lighting load schedules, as a fraction of peak lighting loads, in Figure 4-3 are representative of typical residential occupant activity where the greatest loads are in the late evening between 7:00 PM and 11:00 PM. There is also a spike in lighting loads in the morning between 7:00 AM and 10:00 PM. The lighting loads also vary slightly based on the day of the week.

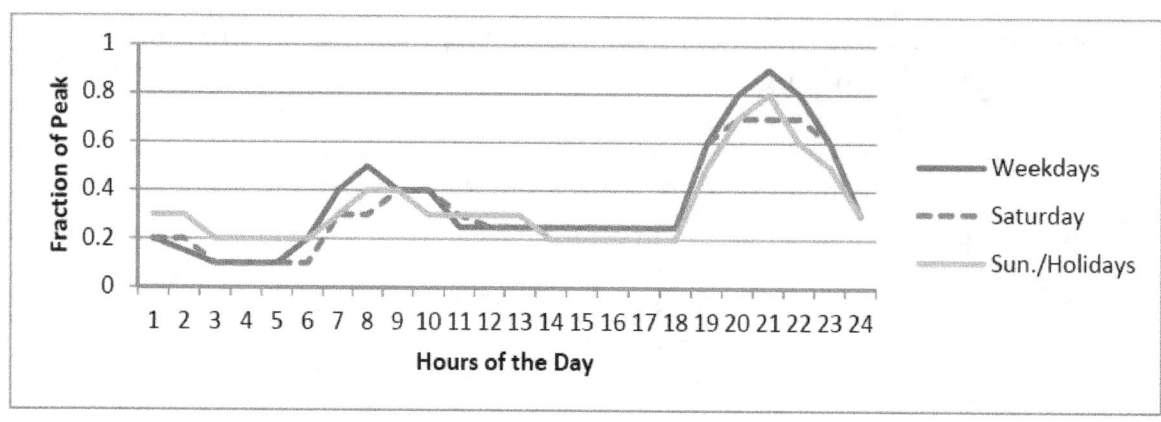

Figure 4-3 APART04 Lighting Schedule

The peak electrical equipment load is 7500 W, or 2.69 W/m^2 (0.25 W/ft^2). Similar to lighting loads, the electrical load schedule in Figure 4-4 is highly correlated with occupant activity.

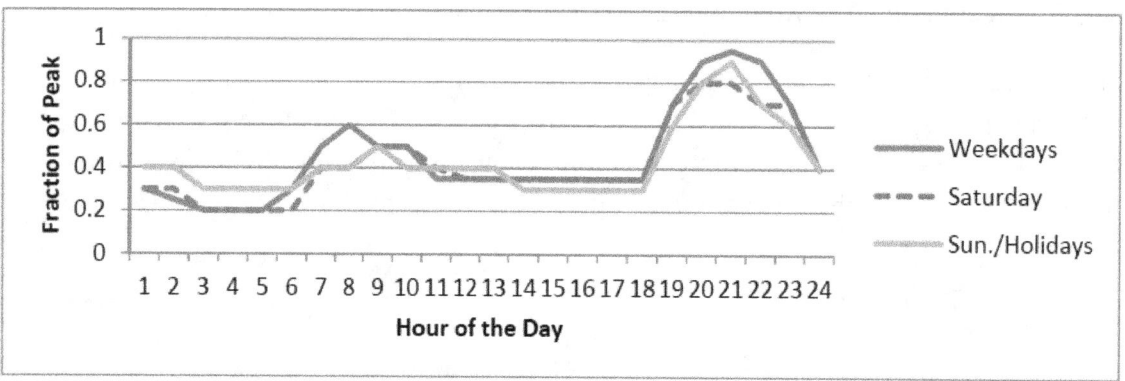

Figure 4-4 APART04 Electrical Load Schedule

4.2   6-Story Apartment Building

The 6-story apartment building has mass walls, insulation entirely above the roof deck, operable windows, and a window-to-wall ratio of 14 %. The detailed assumptions used in the energy simulations are described below.

4.2.1   Building Envelope

The energy efficiency characteristics of the building envelope are determined by the building's location and the edition of *ASHRAE 90.1*. The window characteristics (U-factor, SHGC, and VT) are based on the *ASHRAE 90.1* requirements for operable windows for 10.1 % to 20.0 % glazing. The wall and roof efficiency characteristics are based on the *ASHRAE 90.1* requirements for residential buildings with above grade, mass wall construction and insulation entirely above the roof deck.

14

4.2.2 Heating, Ventilation, and Air Conditioning

There are four main aspects to the heating, ventilation, and air conditioning of a building: equipment, operating conditions, air infiltration, and mechanical ventilation. The HVAC equipment is a packaged electric air-cooled chiller and natural gas-fired hot water boiler. Each building type that falls into the "Lodging" CBECS category has the same constant heating and cooling setpoint temperatures shown in Figure 4-5: 21°C (70°F) for heating and 24°C (76°F) for cooling.

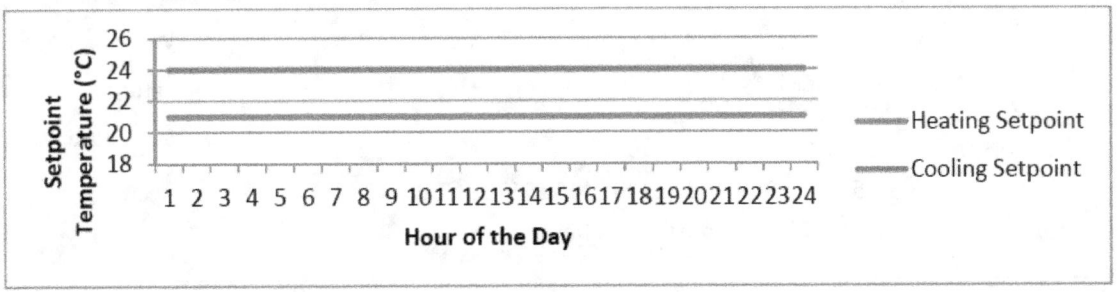

Figure 4-5 APART06 Setpoint Temperature Schedules

Air infiltration and mechanical ventilation are assumed to be constant across all editions of *ASHRAE 90.1*, with an infiltration rate of 0.244 m^3/s per floor (0.30 ACH) and minimum mechanical ventilation of 0.379 m^3/s per floor (0.47 ACH).

4.2.3 Occupancy, Lighting, and Electrical Loads

The peak occupancy for the 6-story apartment building is assumed to be 240 people or 1 person per 23.2 m^2 (250 ft^2). The schedule in Figure 4-6 shows that the greatest occupancy occurs over the nighttime while the lowest occupancy is during the middle of the day.

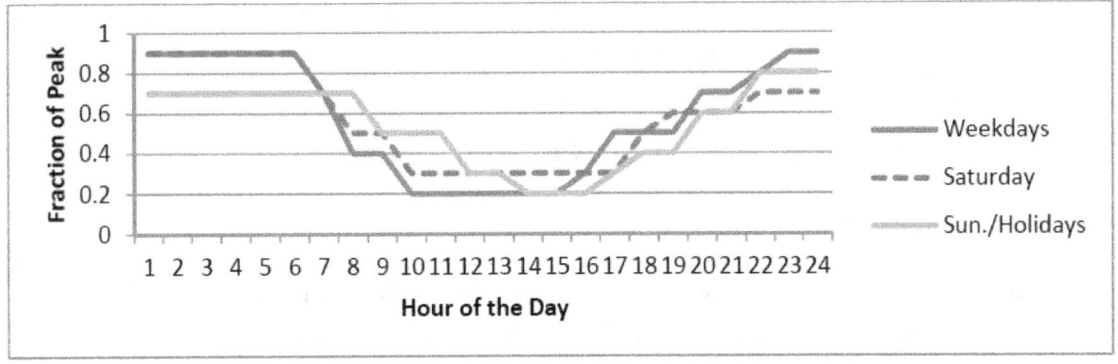

Figure 4-6 APART06 Occupancy Schedule

The energy simulation assumes between 8.6 W/m^2 (0.8 W/ft^2) and 18.3 W/m^2 (1.7 W/ft^2) of lighting density depending on the building design (e.g. edition of *ASHRAE 90.1* or

LEC). The lighting load schedules, as a fraction of peak lighting loads, in Figure 4-7 are representative of typical residential occupant activity where the greatest loads are in the late evening between 7:00 PM and 11:00 PM. There is also a spike in lighting loads in the morning between 7:00 AM and 10:00 PM. The lighting loads also vary slightly based on the day of the week.

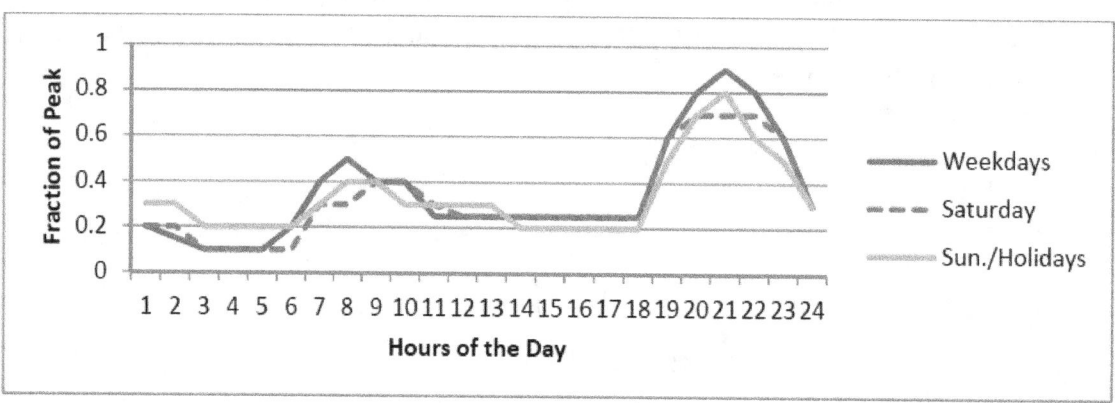

Figure 4-7 APART06 Lighting Schedule

The peak electrical equipment load is 5625 W, or 2.69 W/m^2 (0.25 W/ft^2). Similar to lighting loads, the electrical load schedule in Figure 4-8 is highly correlated with occupant activity.

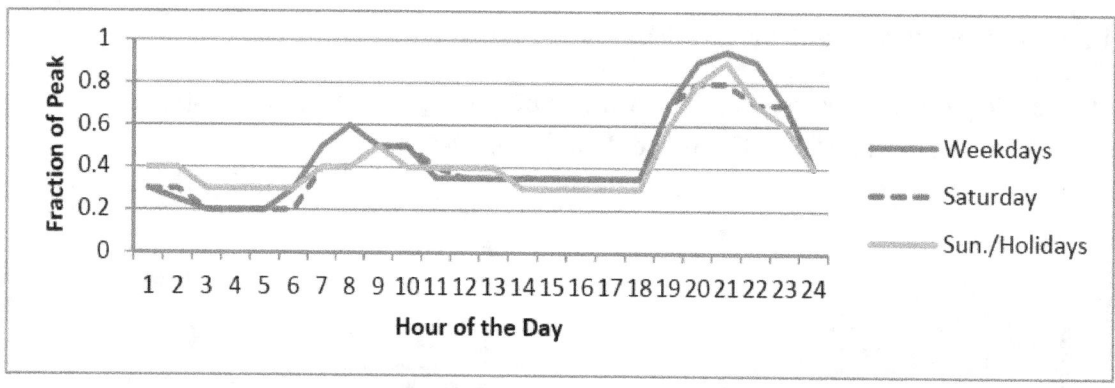

Figure 4-8 APART06 Electrical Load Schedule

4.3 4-Story Dormitory

The 4-story dormitory has mass walls, insulation entirely above the roof deck, operable windows, and a window-to-wall ratio of 20 %. The detailed assumptions used in the energy simulations are described below.

4.3.1 Building Envelope

The energy efficiency characteristics of the building envelope are determined by the building's location and the edition of *ASHRAE 90.1*. The window characteristics (U-factor, SHGC, and VT) are based on the *ASHRAE 90.1* requirements for operable windows for 10.1 % to 20.0 % glazing. The wall and roof efficiency characteristics are based on the *ASHRAE 90.1* requirements for residential buildings with above grade, mass wall construction and insulation entirely above the roof deck.

4.3.2   Heating, Ventilation, and Air Conditioning

There are four main aspects to the heating, ventilation, and air conditioning of a building: equipment, operating conditions, air infiltration, and mechanical ventilation. The HVAC equipment is a packaged electric air-cooled chiller and natural gas-fired hot water boiler. Each building type that falls into the "Lodging" CBECS category has the same constant heating and cooling setpoint temperatures shown in Figure 4-9: 21°C (70°F) for heating and 24°C (76°F) for cooling.

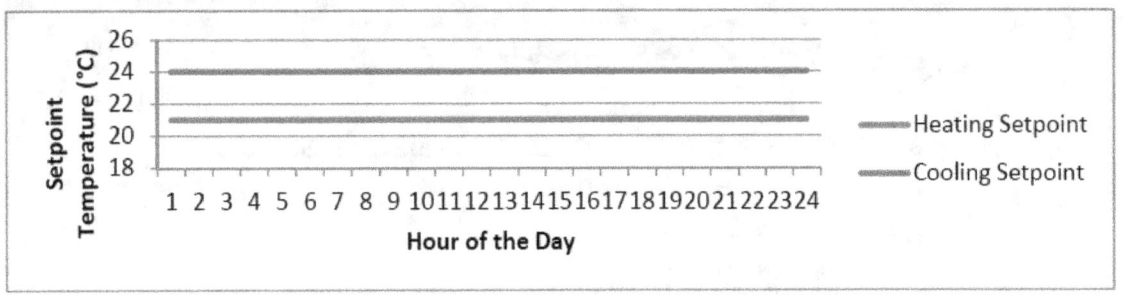

Figure 4-9   DORMI04 Setpoint Temperature Schedules

Air infiltration and mechanical ventilation are assumed to be constant across all editions of *ASHRAE 90.1*, with an infiltration rate of 0.236 m³/s per floor (0.30 ACH) and minimum mechanical ventilation of 0.316 m³/s per floor (0.40 ACH).

4.3.3   Occupancy, Lighting, and Electrical Loads

The peak occupancy for the 3-story dormitory is assumed to be 132 people or 1 person per 23.2 m² (250 ft²). The schedule in Figure 4-10 shows that the greatest occupancy occurs over the nighttime while the lowest occupancy is during the middle of the day.

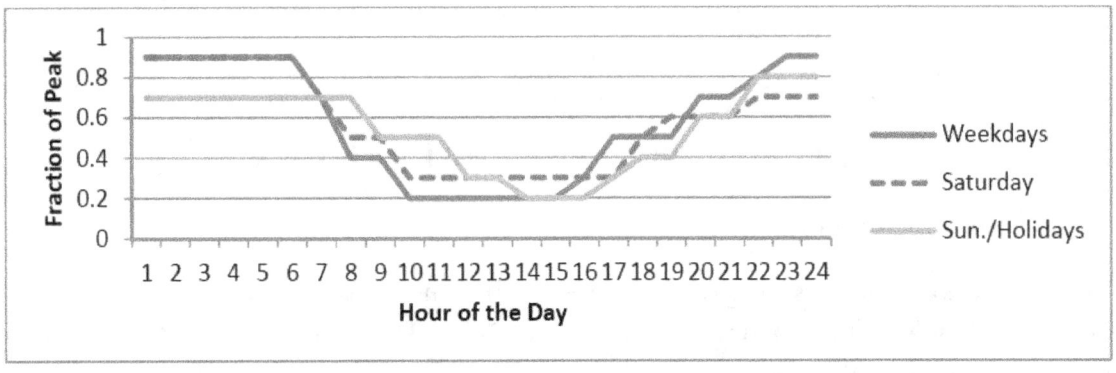

Figure 4-10 DORMI04 Occupancy Schedule

The energy simulation assumes between 8.6 W/m² (0.8 W/ft²) and 18.3 W/m² (1.7 W/ft²) of lighting density depending on the building design (e.g. edition of *ASHRAE 90.1* or LEC). The lighting load schedules, as a fraction of peak lighting loads, in Figure 4-11 are representative of typical residential occupant activity where the greatest loads are in the late evening between 7:00 PM and 11:00 PM. There is also a spike in lighting loads in the morning between 7:00 AM and 10:00 PM. The lighting loads also vary slightly based on the day of the week.

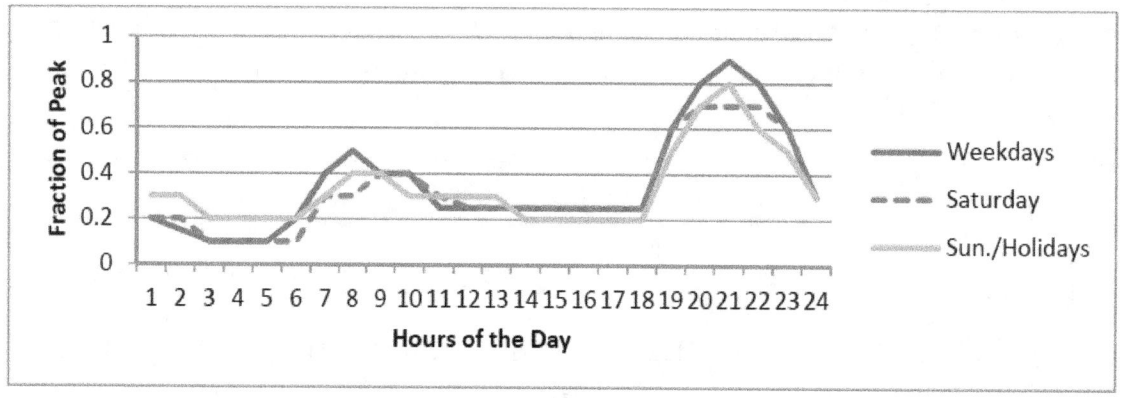

Figure 4-11 DORMI04 Lighting Schedule

The peak electrical equipment load is 8331 W, or 2.69 W/m² (0.25 W/ft²). Similar to lighting loads, the electrical load schedule in Figure 4-12 is highly correlated with occupant activity.

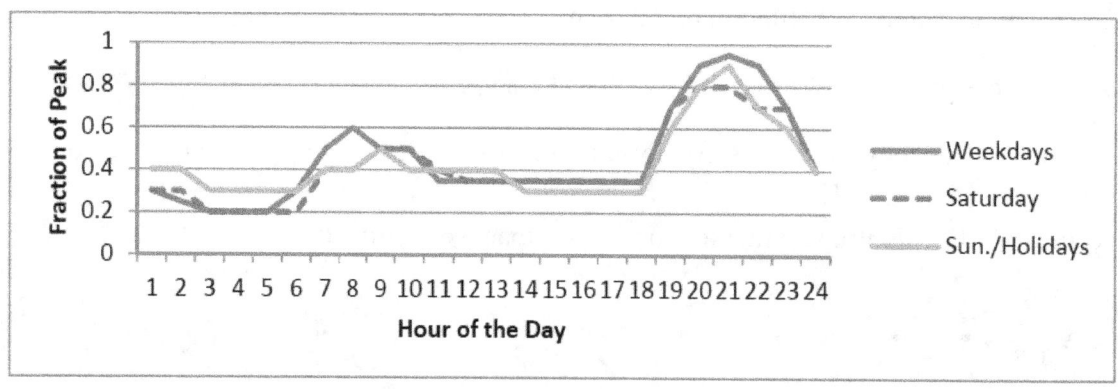

Figure 4-12 DORMI04 Electrical Load Schedule

4.4   6-Story Dormitory

The 6-story dormitory has steel-framed walls, insulation entirely above the roof deck, operable windows, and a window-to-wall ratio of 20 %. The detailed assumptions used in the energy simulations are described below.

4.4.1 Building Envelope

The energy efficiency characteristics of the building envelope are determined by the building's location and the edition of *ASHRAE 90.1*. The window characteristics (U-factor, SHGC, and VT) are based on the *ASHRAE 90.1* requirements for operable windows for 10.1 % to 20.0 % glazing. The wall and roof efficiency characteristics are based on the *ASHRAE 90.1* requirements for residential buildings with above grade, steel-framed wall construction and insulation entirely above the roof deck.

4.4.2 Heating, Ventilation, and Air Conditioning

There are four main aspects to the heating, ventilation, and air conditioning of a building: equipment, operating conditions, air infiltration, and mechanical ventilation. The HVAC equipment is a packaged electric air-cooled chiller and natural gas-fired hot water boiler. Each building type that falls into the "Lodging" CBECS category has the same constant heating and cooling setpoint temperatures shown in Figure 4-13: 21°C (70°F) for heating and 24°C (76°F) for cooling.

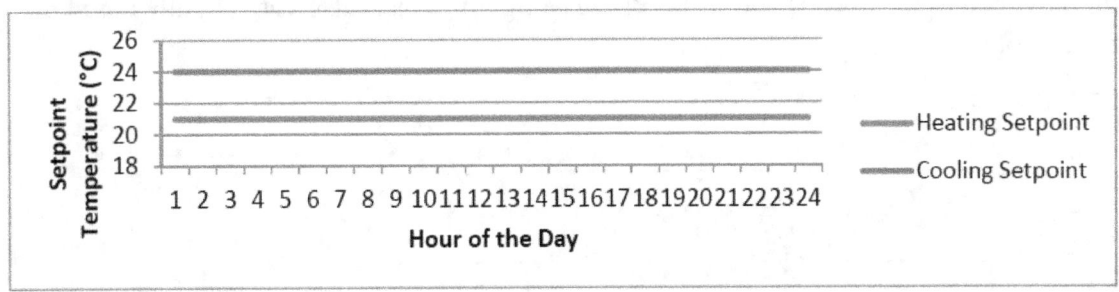

Figure 4-13 DORMI06 Setpoint Temperature Schedules

Air infiltration and mechanical ventilation are assumed to be constant across all editions of *ASHRAE 90.1*, with an infiltration rate of 0.401 m^3/s per floor (0.30 ACH) and minimum mechanical ventilation of 0.537 m^3/s per floor (0.40 ACH).

4.4.3 Occupancy

The peak occupancy for the 6-story dormitory is assumed to be 342 people or 1 person per 23.2 m^2 (250 ft^2). The schedule in Figure 4-14 shows that the greatest occupancy occurs over the nighttime while the lowest occupancy is during the middle of the day.

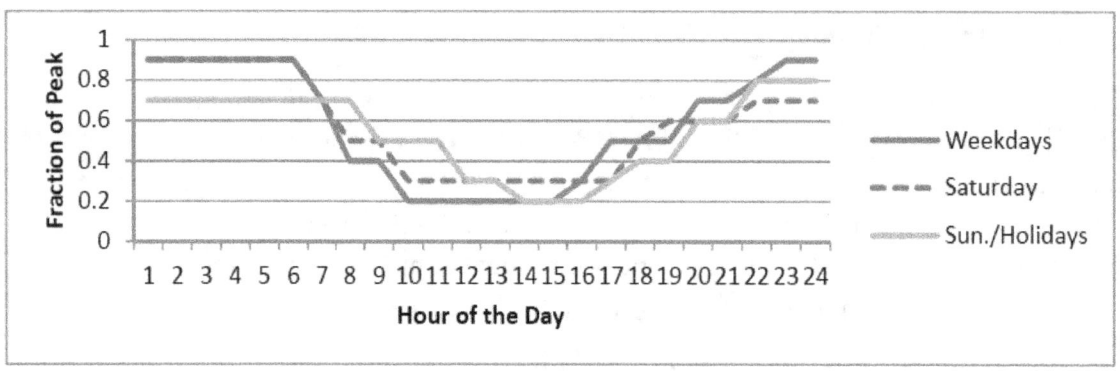

Figure 4-14 DORMI06 Occupancy Schedule

The energy simulation assumes between 8.6 W/m² (0.8 W/ft²) and 18.3 W/m² (1.7 W/ft²) of lighting density depending on the building design (e.g. edition of *ASHRAE 90.1* or LEC). The lighting load schedules, as a fraction of peak lighting loads, in Figure 4-15 are representative of typical residential occupant activity where the greatest loads are in the late evening between 7:00 PM and 11:00 PM. There is also a spike in lighting loads in the morning between 7:00 AM and 10:00 PM. The lighting loads also vary slightly based on the day of the week.

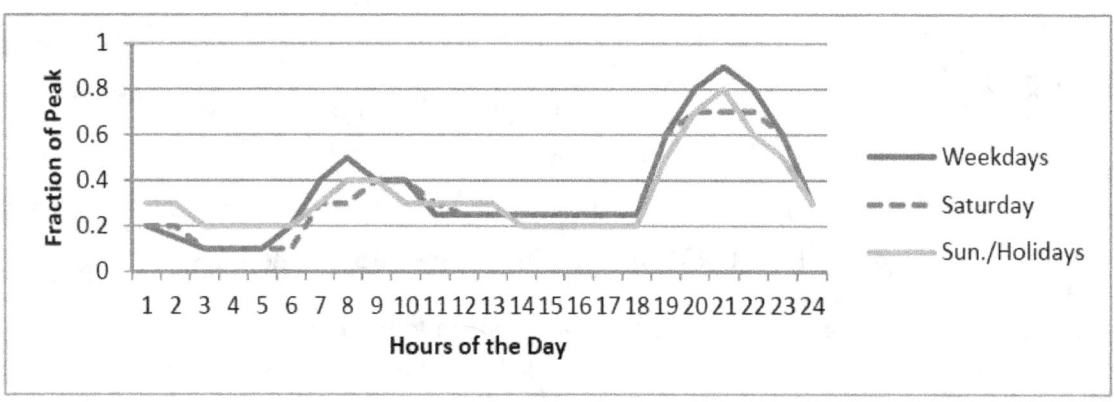

Figure 4-15 DORMI06 Lighting Schedule

The peak electrical equipment load is 21 243 W, or 2.69 W/m² (0.25 W/ft²). Similar to lighting loads, the electrical load schedule in Figure 4-16 is highly correlated with occupant activity.

20

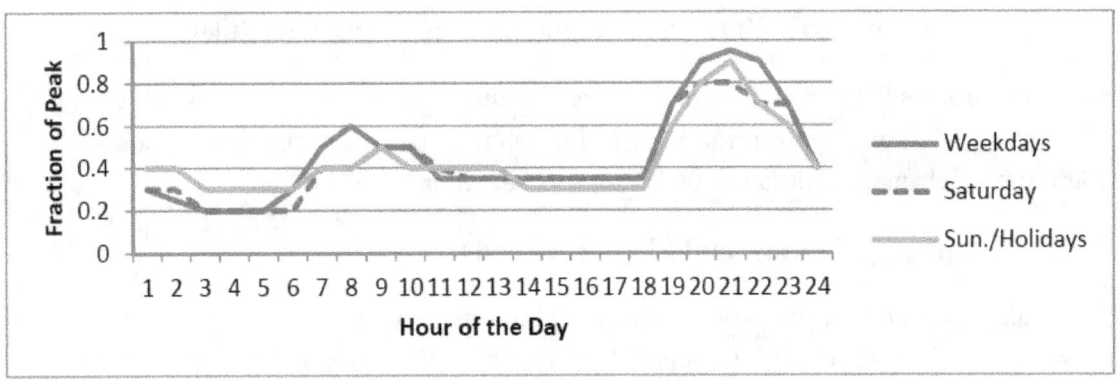

Figure 4-16 DORMI06 Electrical Load Schedule

4.5   15-Story Hotel

The 15-story hotel has glass and metal curtain walls with steel framing, insulation entirely above the roof deck, and a window-to-wall ratio of 100 %. The detailed assumptions used in the energy simulations are described below.

4.5.1   Building Envelope

The energy efficiency characteristics of the building envelope are determined by the building's location and the edition of *ASHRAE 90.1*. The window characteristics (U-factor, SHGC, and VT) are based on the *ASHRAE 90.1* requirements for fixed windows for 40.1 % to 50.0 % glazing. The wall and roof efficiency characteristics are based on the *ASHRAE 90.1* requirements for residential buildings with above grade, steel-framed wall construction and insulation entirely above the roof deck.

4.5.2   Heating, Ventilation, and Air Conditioning

There are four main aspects to the heating, ventilation, and air conditioning of a building: equipment, operating conditions, air infiltration, and mechanical ventilation. The HVAC equipment is a packaged electric water-cooled chiller and natural gas-fired hot water boiler. Each building type that falls into the "Lodging" CBECS category has the same constant heating and cooling setpoint temperatures shown in Figure 4-17: 21°C (70°F) for heating and 24°C (76°F) for cooling.

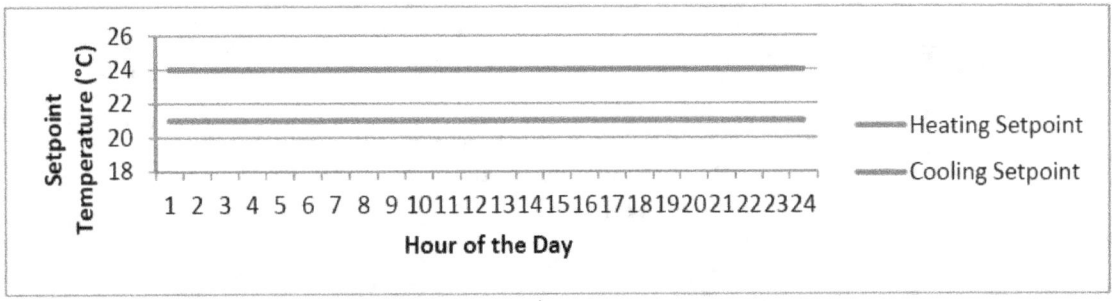

Figure 4-17 HOTEL15 Setpoint Temperature Schedules

Air infiltration and mechanical ventilation are assumed to be constant across all editions of *ASHRAE 90.1*, with an infiltration rate of 0.708 m³/s per floor (0.30 ACH) and minimum mechanical ventilation of 1.136 m³/s per floor (0.48 ACH).

4.5.3 Occupancy, Lighting, and Electrical Loads

The peak occupancy for the 6-story apartment building is assumed to be 1800 people or 1 person per 23.2 m² (250 ft²). The schedule in Figure 4-18 shows that the greatest occupancy occurs over the nighttime while the lowest occupancy is during the middle of the day.

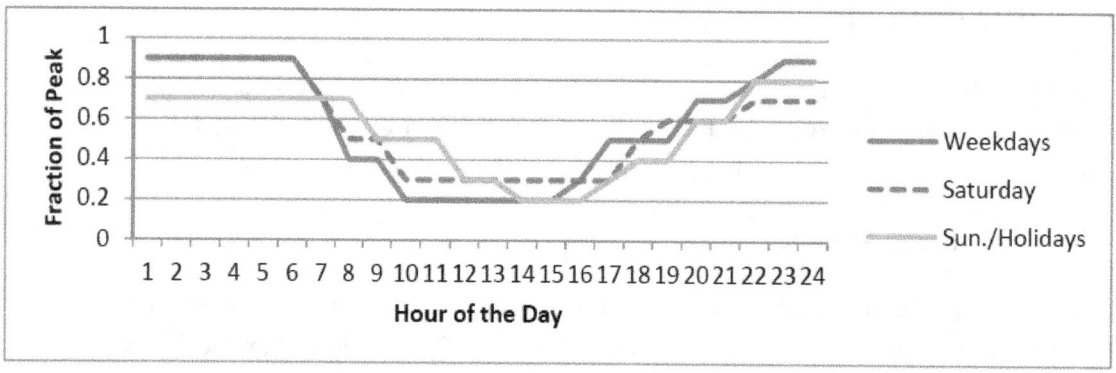

Figure 4-18 HOTEL15 Occupancy Schedule

The energy simulation assumes between 8.6 W/m² (0.8 W/ft²) and 18.3 W/m² (1.7 W/ft²) of lighting density depending on the building design (e.g. edition of *ASHRAE 90.1* or LEC). The lighting load schedules, as a fraction of peak lighting loads, in Figure 4-19 are representative of typical residential occupant activity where the greatest loads are in the late evening between 7:00 PM and 11:00 PM. There is also a spike in lighting loads in the morning between 7:00 AM and 10:00 PM. The lighting loads also vary slightly based on the day of the week.

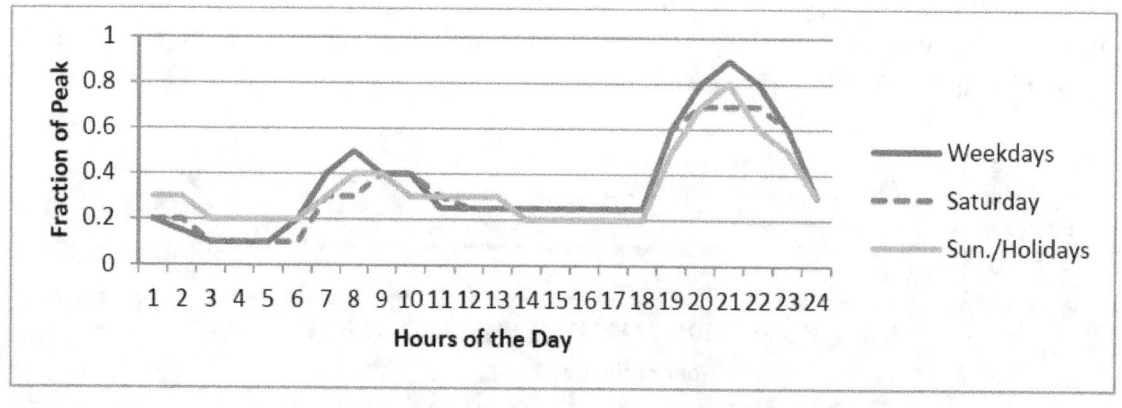

Figure 4-19 HOTEL15 Lighting Schedule

The peak electrical equipment load is 112 462 W, or 2.69 W/m^2 (0.25 W/ft^2). Similar to lighting loads, the electrical load schedule in Figure 4-20 is highly correlated with occupant activity.

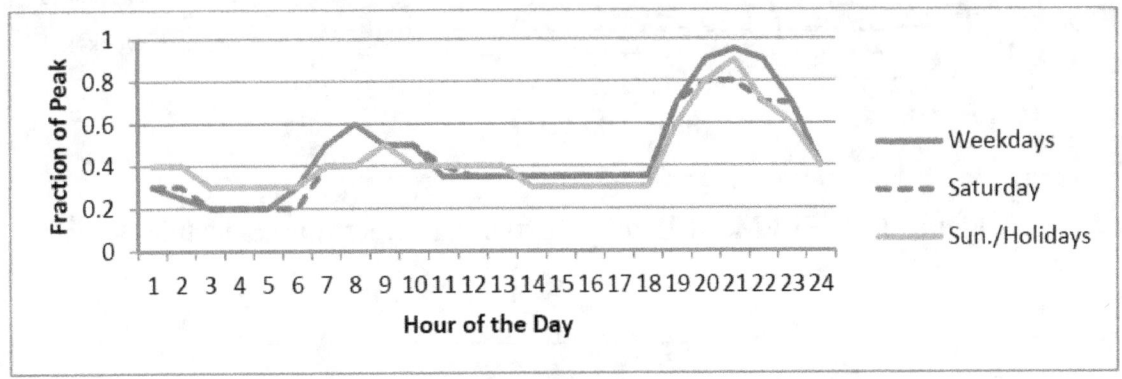

Figure 4-20 HOTEL15 Electrical Load Schedule

4.6   One-Story Elementary School

The one-story elementary school has mass walls, insulation entirely above the roof deck, operable windows, and a window-to-wall ratio of 25 %. The detailed assumptions used in the energy simulations are described below.

4.6.1   Building Envelope

The energy efficiency characteristics of the building envelope are determined by the building's location and the edition of *ASHRAE 90.1*. The window characteristics (U-factor, SHGC, and VT) are based on the *ASHRAE 90.1* requirements for operable windows for 20.1 % to 30.0 % glazing. The wall and roof efficiency characteristics are based on the *ASHRAE 90.1* requirements for nonresidential buildings with above grade, mass wall construction and insulation entirely above the roof deck.

4.6.2   Heating, Ventilation, and Air Conditioning

There are four main aspects to the heating, ventilation, and air conditioning of a building: equipment, operating conditions, air infiltration, and mechanical ventilation. The HVAC equipment is a split system with condensing unit and a natural gas-fired hot water boiler. Each building type that falls into the "Education" CBECS category has the same heating and cooling setpoint temperature schedules. Figure 4-21 shows that the heating setpoint temperature is a constant 16°C (61°F) for weekends while it is 21°C (70°F) during the daytime and 16°C (61°F) during the nighttime on weekdays. Similarly, Figure 4-22 shows that the cooling setpoint temperature is a constant 31°C (88°F) for weekends while it is 25°C (77°F) during the daytime and 31°C (88°F) during the nighttime weekdays. These setpoints correlate with the building occupancy schedule.

23

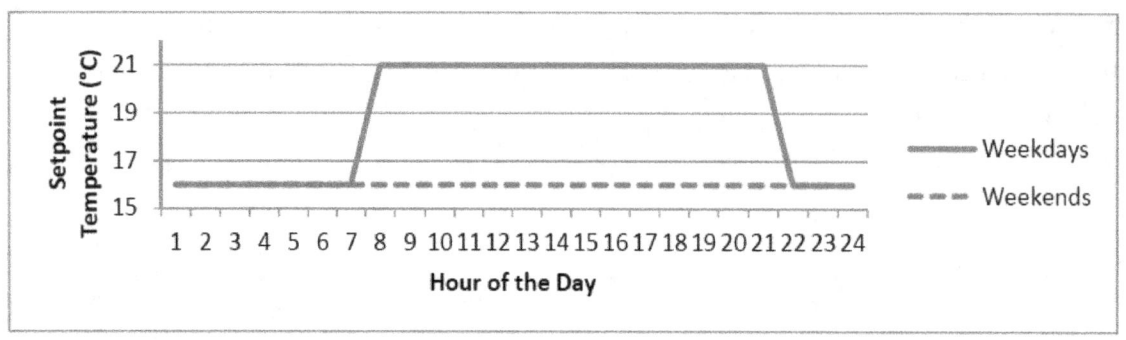

Figure 4-21 ELEMS01 Heating Setpoint Temperature Schedule

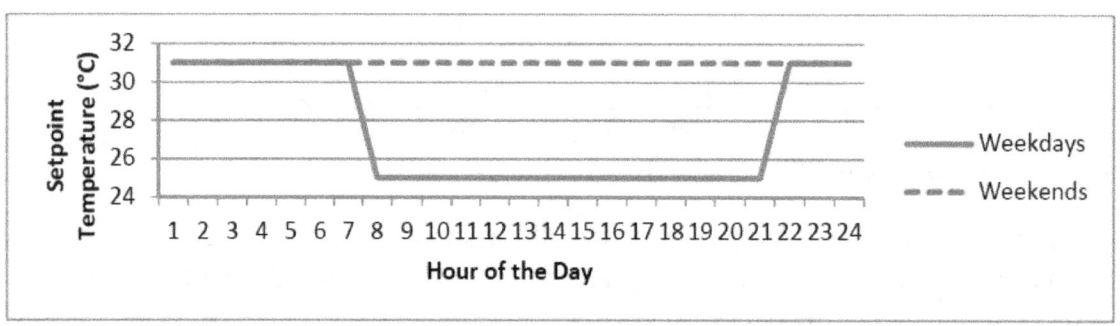

Figure 4-22 ELEMS01 Cooling Setpoint Temperature Schedule

Air infiltration and mechanical ventilation are assumed to be constant across all editions of *ASHRAE 90.1*, with an infiltration rate of 1.593 m^3/s per floor (0.30 ACH) and minimum mechanical ventilation of 5.519 m^3/s per floor (1.04 ACH).

4.6.3   Occupancy, Lighting, and Electrical Loads

The peak occupancy for the 1-story elementary school is assumed to be 602 people or 1 person per 7.0 m^2 (75 ft^2). The schedule in Figure 4-23 shows that the greatest occupancy occurs on the weekdays during the school year. The elementary school is assumed to not be used during the nighttime or on weekends year-round. On weekdays during the summer, the occupancy is much lower than during the school year.

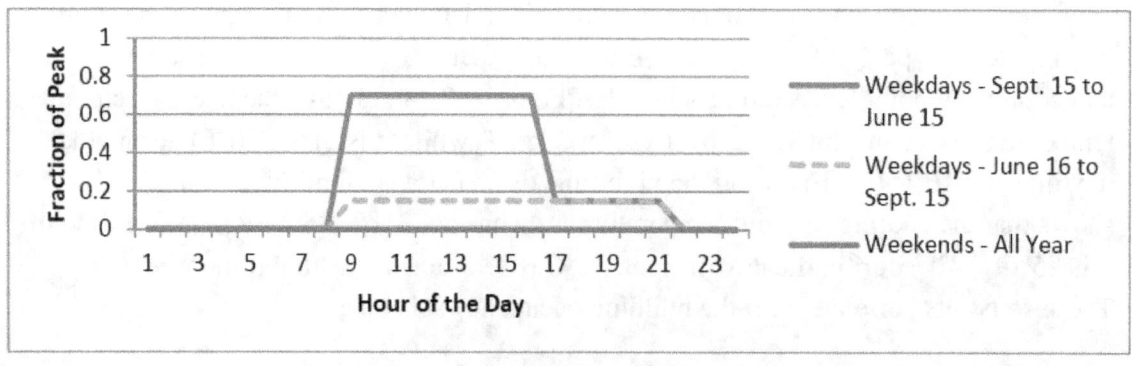

Figure 4-23 ELEMS01 Occupancy Schedule

The energy simulation assumes between 10.8 W/m^2 (1.0 W/ft^2) and 16.1 W/m^2 (1.5 W/ft^2) of lighting density depending on the building design (e.g. edition of *ASHRAE 90.1* or LEC). The lighting load schedules, as a fraction of peak lighting loads, in Figure 4-24 are representative of typical school occupant activity. The loads are greatest during daytime hours on weekdays of the school year. Daytime loads are lower during the summer while school is out. There is no lighting use during the nighttime or on weekends year-round.

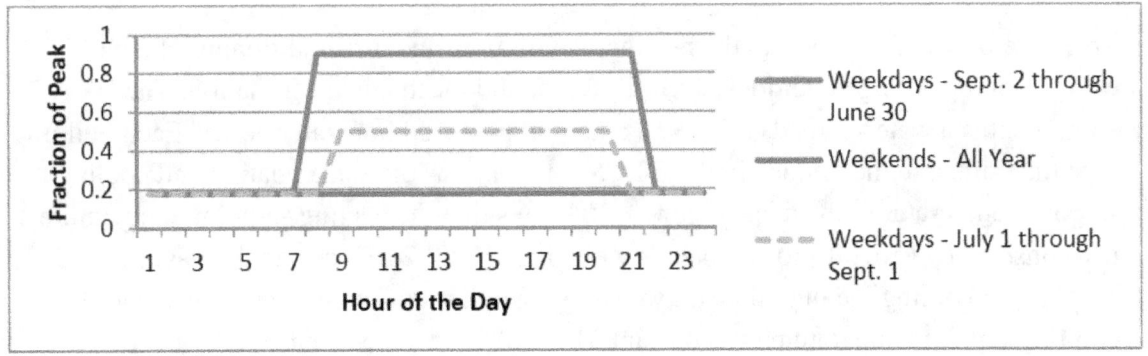

Figure 4-24 ELEMS01 Lighting Schedule

The peak electrical equipment load is 32 489 W, or 5.38 W/m^2 (0.5 W/ft^2). The electrical load schedule in Figure 4-25 is highly correlated with the times of the year that children and teachers are at the school.

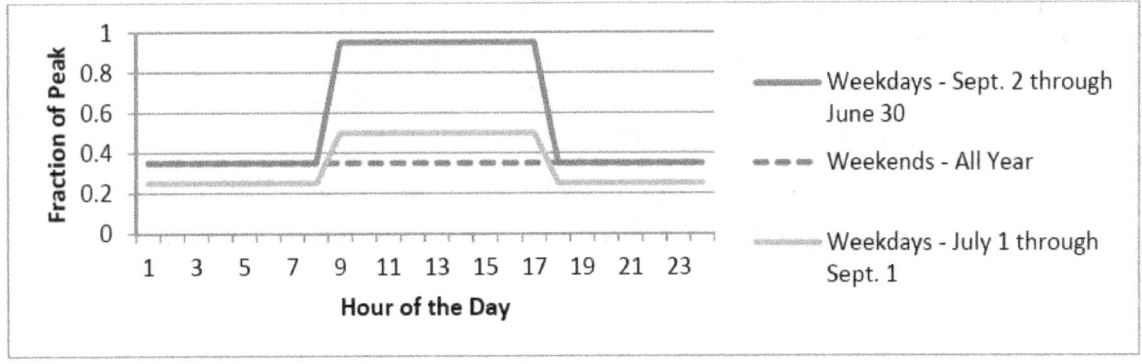

Figure 4-25 ELEMS01 Electrical Load Schedule

4.7 2-Story High School

The 2-story high school has mass walls, insulation entirely above the roof deck, operable windows, and a window-to-wall ratio of 25 %. The detailed assumptions used in the energy simulations are described below.

4.7.1   Building Envelope

The energy efficiency characteristics of the building envelope are determined by the building's location and the edition of *ASHRAE 90.1*. The window characteristics (U-factor, SHGC, and VT) are based on the *ASHRAE 90.1* requirements for operable windows for 20.1 % to 30.0 % glazing. The wall and roof efficiency characteristics are based on the *ASHRAE 90.1* requirements for nonresidential buildings with above grade, mass wall construction and insulation entirely above the roof deck.

4.7.2   Heating, Ventilation, and Air Conditioning

There are four main aspects to the heating, ventilation, and air conditioning of a building: equipment, operating conditions, air infiltration, and mechanical ventilation. The HVAC equipment is a water-cooled chiller and a natural gas-fired hot water boiler. Each building type that falls into the "Education" CBECS category has the same heating and cooling setpoint temperature schedules. Figure 4-26 shows that the heating setpoint temperature is a constant 16°C (61°F) for weekends while it is 21°C (70°F) during the daytime and 16°C (61°F) during the nighttime on weekdays. Similarly, Figure 4-27 shows that the cooling setpoint temperature is a constant 31°C (88°F) for weekends while it is 25°C (77°F) during the daytime and 31°C (88°F) during the nighttime weekdays. These setpoints correlate with the building occupancy schedule.

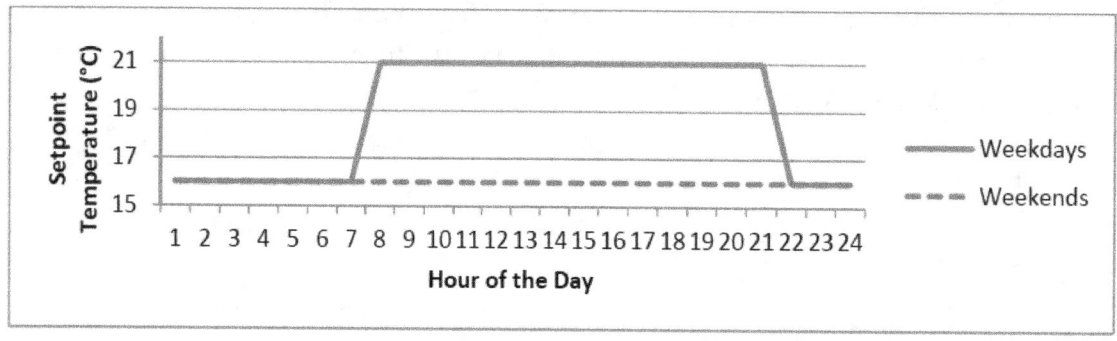

Figure 4-26 HIGHS02 Heating Setpoint Temperature Schedule

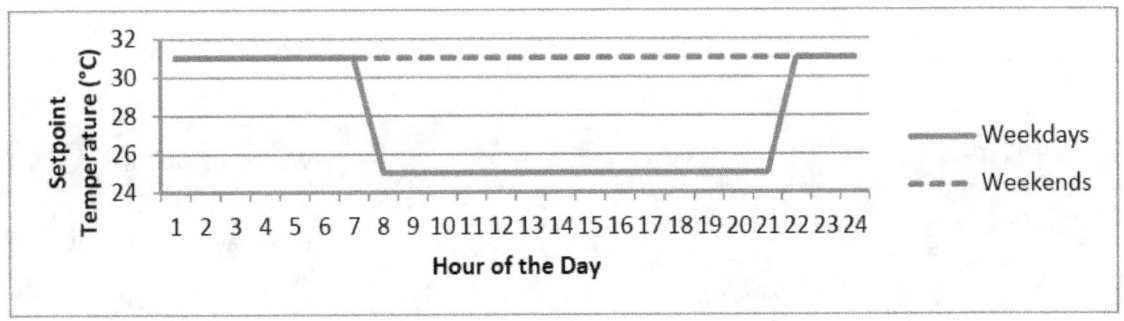

Figure 4-27 HIGHS02 Cooling Setpoint Temperature Schedule

Air infiltration and mechanical ventilation are assumed to be constant across all editions of *ASHRAE 90.1*, with an infiltration rate of 2.301 m³/s per floor (0.30 ACH) and minimum mechanical ventilation of 7.971 m³/s per floor (1.04 ACH).

4.7.3 Occupancy, Lighting, and Electrical Loads

The peak occupancy for the 2-story high school is assumed to be 1740 people or 1 person per 7.0 m² (75 ft²). The schedule in Figure 4-28 shows that the greatest occupancy occurs on the weekdays during the school year. The high school is assumed to not be used during the nighttime or on weekends year-round. On weekdays during the summer, the occupancy is much lower than during the school year.

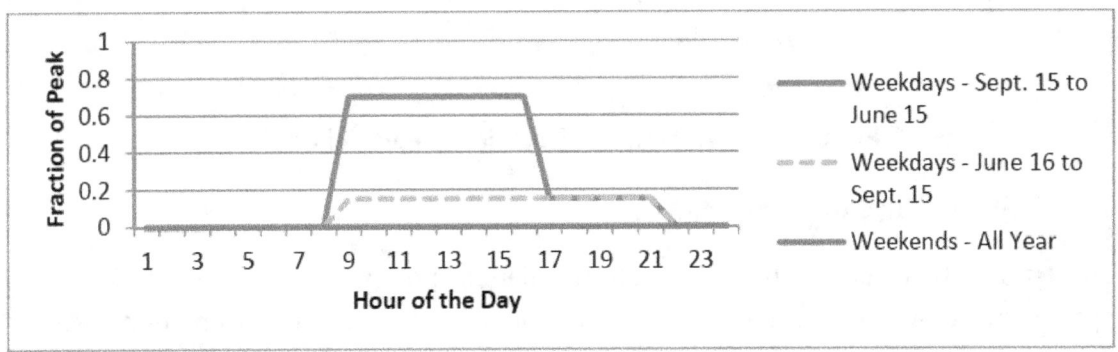

Figure 4-28 HIGHS02 Occupancy Schedule

The energy simulation assumes between 10.8 W/m² (1.0 W/ft²) and 16.1 W/m² (1.5 W/ft²) of lighting density depending on the building design (e.g. edition of *ASHRAE 90.1* or LEC). The lighting load schedules, as a fraction of peak lighting loads, in Figure 4-29 are representative of typical school occupant activity. The loads are greatest during daytime hours on weekdays of the school year. Daytime loads are lower during the summer while school is out. There is no lighting use during the nighttime or on weekends year-round.

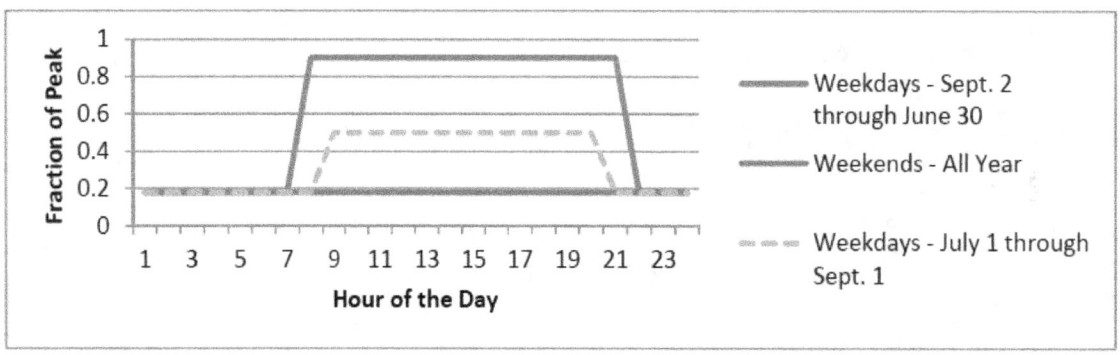

Figure 4-29 HIGHS02 Lighting Schedule

The peak electrical equipment load is 64 978 W, or 5.38 W/m² (0.5 W/ft²). The electrical load schedule in Figure 4-30 is highly correlated with the times of the year that children and teachers are at the school (9 AM to 5 PM).

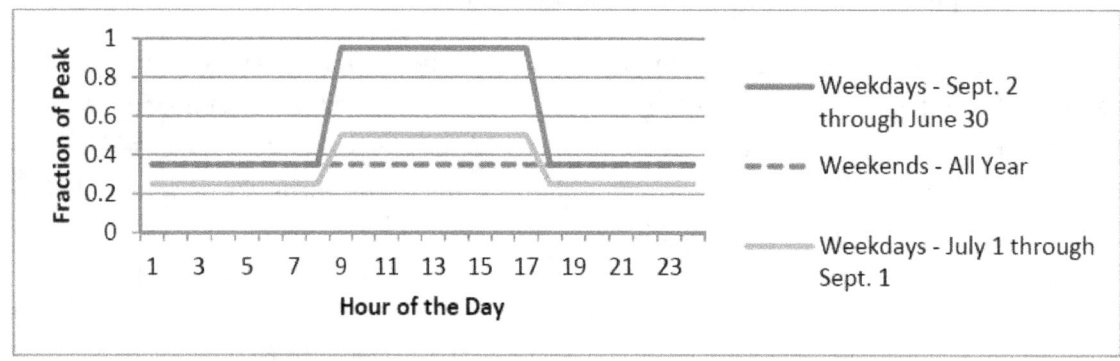

Figure 4-30 HIGHS02 Electrical Load Schedule

4.8 3-Story Office Building

The 3-story office building has mass walls, insulation entirely above the roof deck, operable windows, and a window-to-wall ratio of 20 %. The detailed assumptions used in the energy simulations are described below.

4.8.1 Building Envelope

The energy efficiency characteristics of the building envelope are determined by the building's location and the edition of *ASHRAE 90.1*. The window characteristics (U-factor, SHGC, and VT) are based on the *ASHRAE 90.1* requirements for operable windows for 10.1 % to 20.0 % glazing. The wall and roof efficiency characteristics are based on the *ASHRAE 90.1* requirements for nonresidential buildings with above grade, mass wall construction and insulation entirely above the roof deck.

4.8.2 Heating, Ventilation, and Air Conditioning

There are four main aspects to the heating, ventilation, and air conditioning of a building: equipment, operating conditions, air infiltration, and mechanical ventilation. The HVAC equipment is an air-cooled electric chiller and a natural gas-fired hot water boiler. Each building type that falls into the "Office" CBECS category has the same heating and cooling setpoint temperature schedules. Figure 4-31 shows that the heating setpoint temperature varies by day of the week. The setpoint is a constant 15.6°C (60°F) for Sundays and holidays while it is 21°C (70°F) from 7 AM to 5 PM and 15.6°C (60°F) for the rest of the day on Saturdays. Weekdays have a similar schedule to Saturdays, with a setpoint of 21°C (70°F) from 6 AM to 7 PM and 15.6°C (60°F) for all other hours. Similarly, Figure 4-32 shows that the cooling setpoint temperature is a constant 30°C (86°F) on Sundays and holidays while it is 24°C (75°F) from 7 AM to 6 PM and 30°C (86°F) the remainder of the day on Saturdays. Weekdays have a setpoint of 24°C (75°F) from 7 AM to 10 PM and 30°C (86°F) for all other hours. These setpoints correlate with the building occupancy schedule.

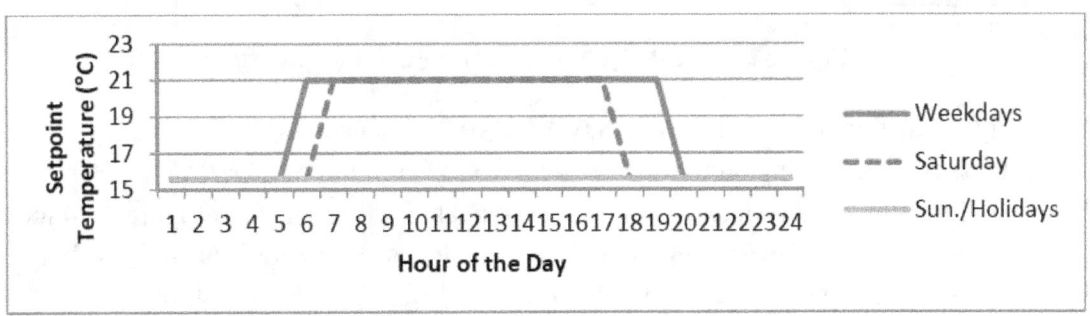

Figure 4-31 OFFIC03 Heating Setpoint Temperature Schedule

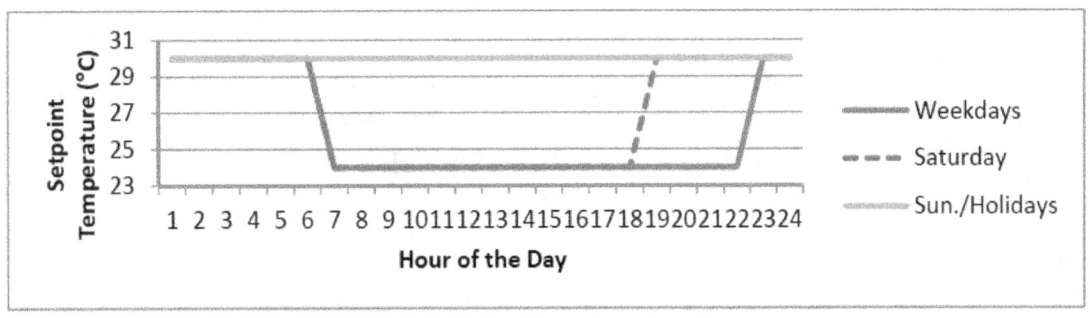

Figure 4-32 OFFIC03 Cooling Setpoint Temperature Schedule

Air infiltration and mechanical ventilation are assumed to be constant across all editions of *ASHRAE 90.1*, with an infiltration rate of 0.189 m³/s per floor (0.30 ACH) and minimum mechanical ventilation of 0.246 m³/s per floor (0.39 ACH).

4.8.3 Occupancy, Lighting, and Electrical Loads

The peak occupancy for the 3-story office building is assumed to be 72 people or 1 person per 25.5 m^2 (275 ft^2). The schedule in Figure 4-33 shows that occupancy represents typical office activity, where the majority of people are in the building during typical "office hours" (9 AM to 5 PM) on the weekdays with a drop in occupancy over the lunch hour. There is a relatively small amount of occupant activity on Saturdays during typical "office hours." There is no occupancy on Sundays or holidays.

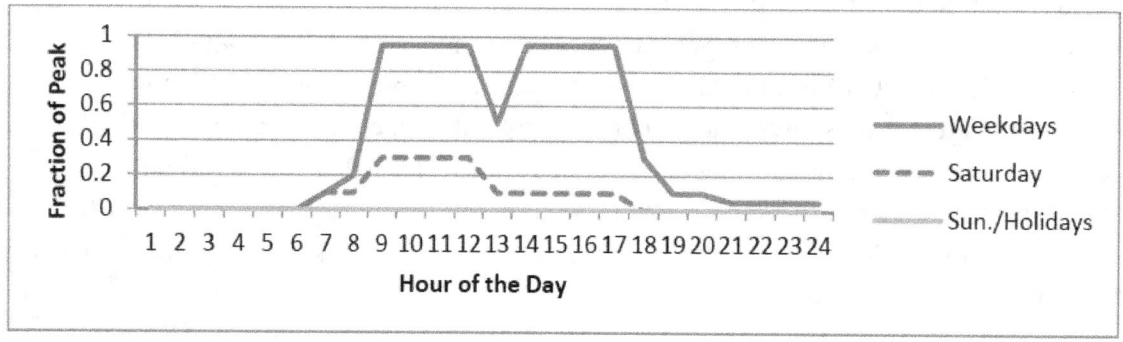

Figure 4-33 OFFIC03 Occupancy Schedule

The energy simulation assumes between 8.6 W/m^2 (0.8W/ft^2) and 14.0 W/m^2 (1.3 W/ft^2) of lighting density depending on the building design (e.g. edition of *ASHRAE 90.1* or LEC). The lighting load schedules, as a fraction of peak lighting loads, in Figure 4-34 are representative of typical office occupant activity. The loads are greatest during "working" hours on weekdays. Daytime loads are lower on Saturdays. There is no lighting use during the nighttime or on Sundays or holidays.

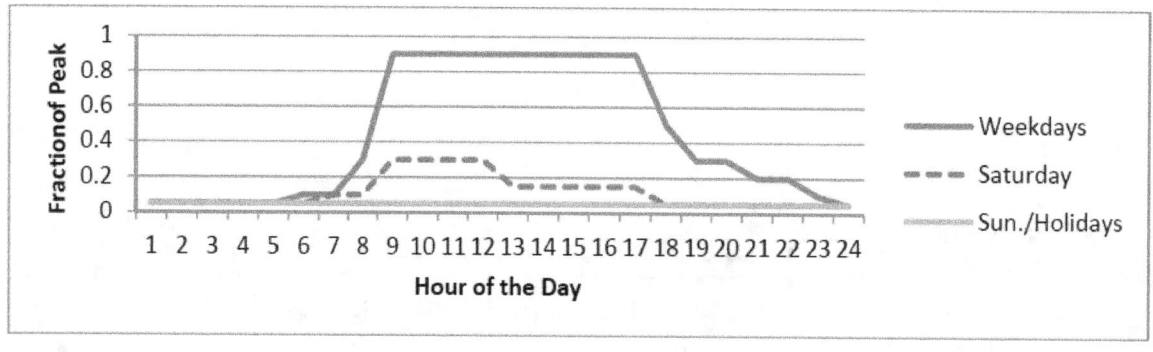

Figure 4-34 OFFIC03 Lighting Schedule

The peak electrical equipment load is 14 996 W, or 8.07 W/m^2 (0.75 W/ft^2). The electrical load schedule in Figure 4-35 is highly correlated with the occupancy schedule with the greatest electrical loads between 9 AM to 5 PM on weekdays.

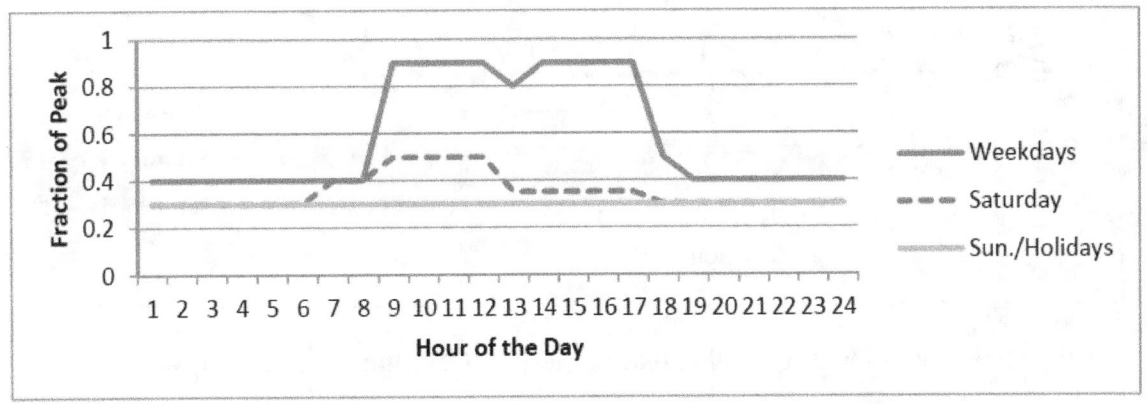

Figure 4-35 OFFIC03 Electrical Load Schedule

4.9   8-Story Office Building

The 8-story office building has mass walls, insulation entirely above the roof deck, operable windows, and a window-to-wall ratio of 20 %. The detailed assumptions used in the energy simulations are described below.

4.9.1   Building Envelope

The energy efficiency characteristics of the building envelope are determined by the building's location and the edition of *ASHRAE 90.1*. The window characteristics (U-factor, SHGC, and VT) are based on the *ASHRAE 90.1* requirements for operable windows for 10.1 % to 20.0 % glazing. The wall and roof efficiency characteristics are based on the *ASHRAE 90.1* requirements for nonresidential buildings with above grade, mass wall construction and insulation entirely above the roof deck.

4.9.2   Heating, Ventilation, and Air Conditioning

There are four main aspects to the heating, ventilation, and air conditioning of a building: equipment, operating conditions, air infiltration, and mechanical ventilation. The HVAC equipment is a rooftop packaged air conditioner and a natural gas-fired furnace. Each building type that falls into the "Office" CBECS category has the same heating and cooling setpoint temperature schedules. Figure 4-36 shows that the heating setpoint temperature varies by day of the week. The setpoint is a constant 15.6°C (60°F) for Sundays and holidays while it is 21°C (70°F) from 7 AM to 5 PM and 15.6°C (60°F) for the rest of the day on Saturdays. Weekdays have a similar schedule to Saturdays, with a setpoint of 21°C (70°F) from 6 AM to 7 PM and 15.6°C (60°F) for all other hours. Similarly, Figure 4-37 shows that the cooling setpoint temperature is a constant 30C on Sundays and holidays while it is 24°C (75°F) from 7 AM to 6 PM and 30°C (86°F) the remainder of the day on Saturdays. Weekdays have a setpoint of 24°C (75°F) from 7 AM to 10 PM and 30°C (86°F) for all other hours. These setpoints correlate with the building occupancy schedule.

31

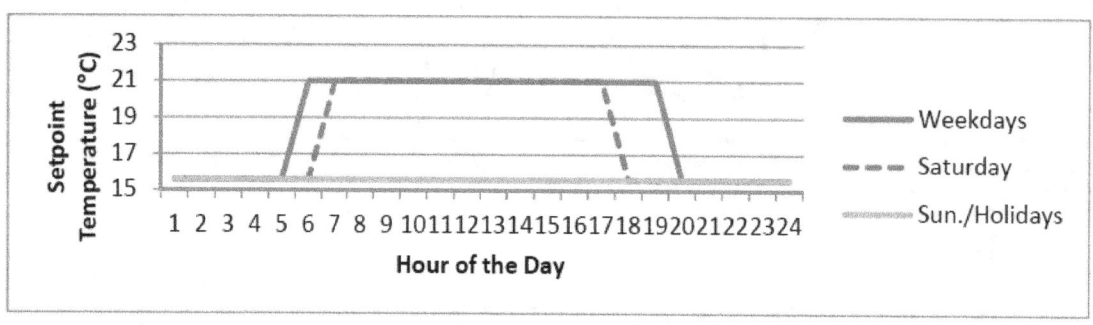

Figure 4-36 OFFIC08 Heating Setpoint Temperature Schedule

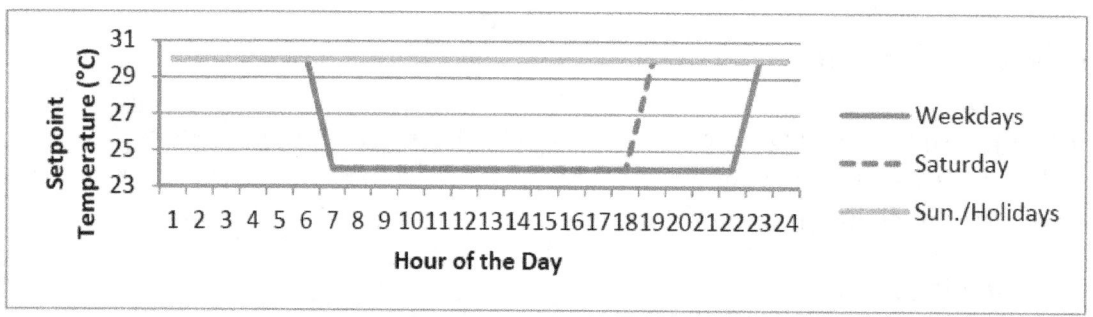

Figure 4-37 OFFIC08 Cooling Setpoint Temperature Schedule

4.9.3 Air Infiltration and Mechanical Ventilation

Air infiltration and mechanical ventilation are assumed to be constant across all editions of *ASHRAE 90.1*, with an infiltration rate of 0.283 m^3/s per floor (0.30 ACH) and minimum mechanical ventilation of 0.370 m^3/s per floor (0.39 ACH).

4.9.4 Occupancy, Lighting, and Electrical Loads

The peak occupancy for the 8-story office building is assumed to be 288 people or 1 person per 25.5 m^2 (275 ft^2). The schedule in Figure 4-38 shows that occupancy represents typical office activity, where the majority of people are in the building during typical "office hours" (9 AM to 5 PM) on the weekdays with a drop in occupancy over the lunch hour. There is a relatively small amount of occupant activity on Saturdays during typical "office hours." There is no occupancy on Sundays or holidays.

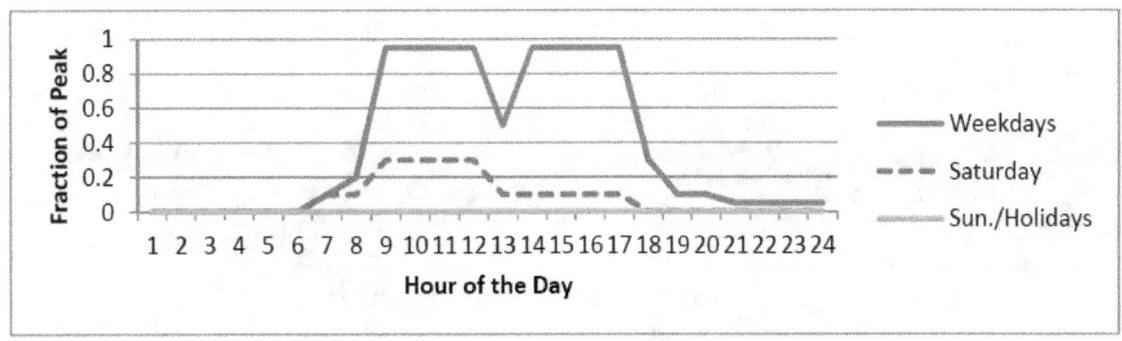

Figure 4-38 OFFIC08 Occupancy Schedule

The energy simulation assumes between 8.6 W/m^2 (0.8W/ft^2) and 14.0 W/m^2 (1.3 W/ft^2) of lighting density depending on the building design (e.g. edition of *ASHRAE 90.1* or LEC). The lighting load schedules, as a fraction of peak lighting loads, in Figure 4-39 are representative of typical office occupant activity. The loads are greatest during "working" hours on weekdays. Daytime loads are lower on Saturdays. There is no lighting use during the nighttime or on Sundays or holidays.

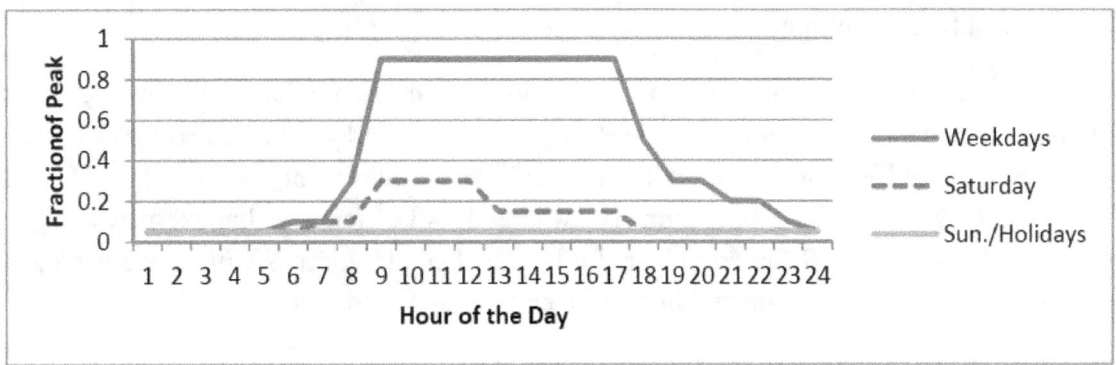

Figure 4-39 OFFIC08 Lighting Schedule

The peak electrical equipment load is 59 978 W, or 8.07 W/m^2 (0.75 W/ft^2). The electrical load schedule in Figure 4-40 is highly correlated with the occupancy schedule with the greatest electrical loads between 9 AM to 5 PM on weekdays.

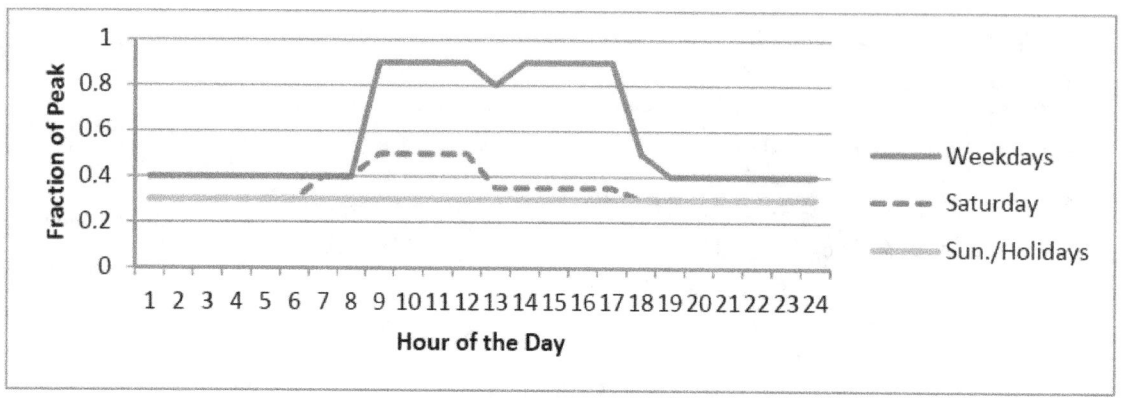

Figure 4-40 OFFIC08 Electrical Load Schedule

4.10 16-Story Office Building

The 16-story office building has glass and metal curtain walls with steel framing, insulation entirely above the roof deck, and a window-to-wall ratio of 100 %. The detailed assumptions used in the energy simulations are described below.

4.10.1 Building Envelope

The energy efficiency characteristics of the building envelope are determined by the building's location and the edition of *ASHRAE 90.1*. The window characteristics (U-factor, SHGC, and VT) are based on the *ASHRAE 90.1* requirements for operable windows for 40.1 % to 50.0 % glazing. The wall and roof efficiency characteristics are based on the *ASHRAE 90.1* requirements for nonresidential buildings with above grade, steel-framed wall construction and insulation entirely above the roof deck.

4.10.2 Heating, Ventilation, and Air Conditioning

There are four main aspects to the heating, ventilation, and air conditioning of a building: equipment, operating conditions, air infiltration, and mechanical ventilation. The HVAC equipment is a water-cooled electric chiller and a natural gas-fired hot-water boiler. Each building type that falls into the "Office" CBECS category has the same heating and cooling setpoint temperature schedules. Figure 4-41 shows that the heating setpoint temperature varies by day of the week. The setpoint is a constant 15.6°C (60°F) for Sundays and holidays while it is 21°C (70°F) from 7 AM to 5 PM and 15.6°C (60°F) for the rest of the day on Saturdays. Weekdays have a similar schedule to Saturdays, with a setpoint of 21°C (70°F) from 6 AM to 7 PM and 15.6°C (60°F) for all other hours. Similarly, Figure 4-42 shows that the cooling setpoint temperature is a constant 30°C (86°F) on Sundays and holidays while it is 24°C (75°F) from 7 AM to 6 PM and 30°C (86°F) the remainder of the day on Saturdays. Weekdays have a setpoint of 24°C (75°F) from 7 AM to 10 PM and 30°C (86°F) for all other hours. These setpoints correlate with the building occupancy schedule.

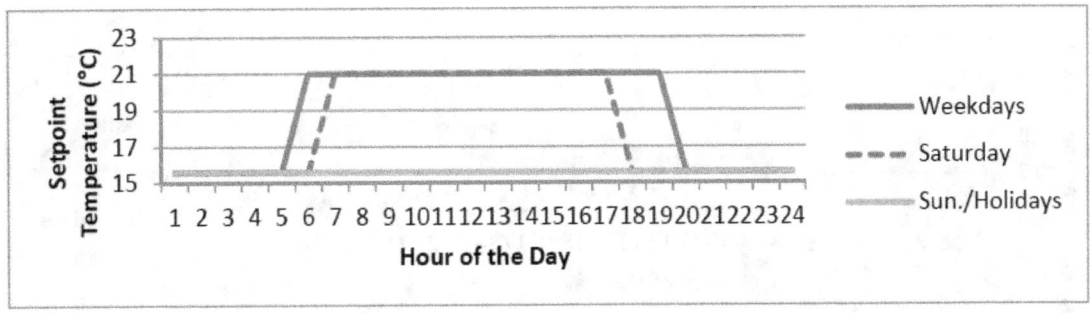

Figure 4-41 OFFIC16 Heating Setpoint Temperature Schedule

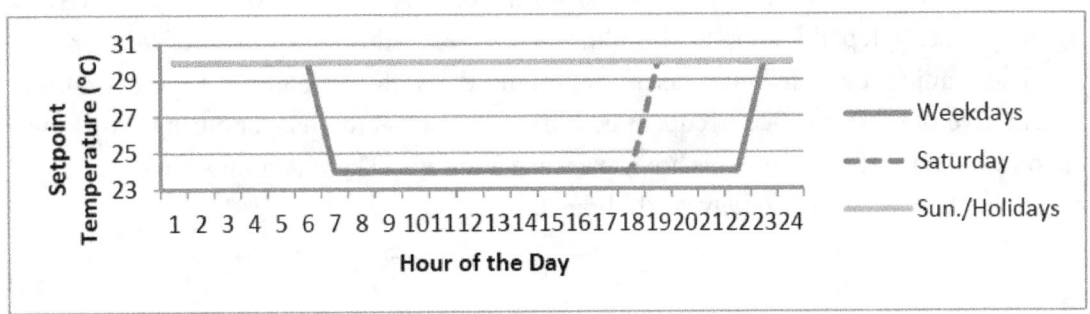

Figure 4-42 OFFIC16 Cooling Setpoint Temperature Schedule

Air infiltration and mechanical ventilation are assumed to be constant across all editions of *ASHRAE 90.1*, with an infiltration rate of 0.384 m^3/s per floor (0.30 ACH) and minimum mechanical ventilation of 0.600 m^3/s per floor (0.47 ACH).

4.10.3 Occupancy, Lighting, and Electrical Loads

The peak occupancy for the 16-story office building is assumed to be 944 people or 1 person per 25.5 m^2 (275 ft^2). The schedule in Figure 4-43 shows that occupancy represents typical office activity, where the majority of people are in the building during typical "office hours" (9 AM to 5 PM) on the weekdays with a drop in occupancy over the lunch hour. There is a relatively small amount of occupant activity on Saturdays during typical "office hours." There is no occupancy on Sundays or holidays.

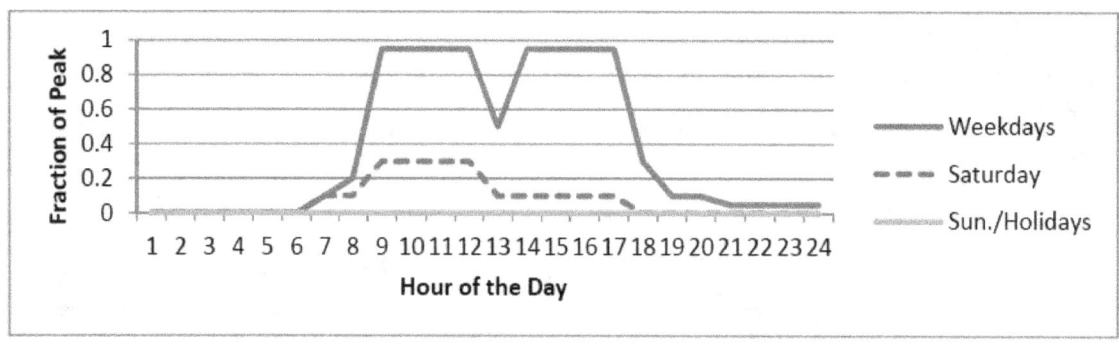

Figure 4-43 OFFIC16 Occupancy Schedule

The energy simulation assumes between 8.6 W/m^2 (0.8W/ft^2) and 14.0 W/m^2 (1.3 W/ft^2) of lighting density depending on the building design (e.g. edition of *ASHRAE 90.1* or LEC). The lighting load schedules, as a fraction of peak lighting loads, in Figure 4-44 are representative of typical office occupant activity. The loads are greatest during "working" hours on weekdays. Daytime loads are lower on Saturdays. There is no lighting use during the nighttime or on Sundays or holidays.

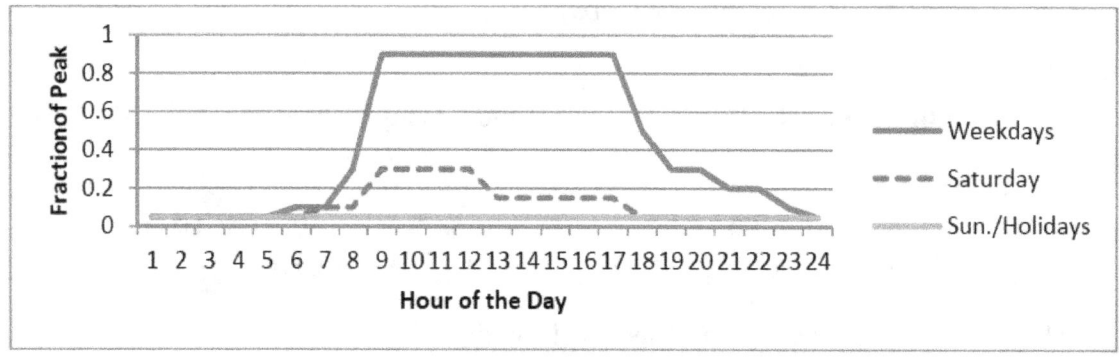

Figure 4-44 OFFIC16 Lighting Schedule

The peak electrical equipment load is 194 934 W, or 8.07 W/m^2 (0.75 W/ft^2). The electrical load schedule in Figure 4-45 is highly correlated with the occupancy schedule with the greatest electrical loads between 9 AM to 5 PM on weekdays.

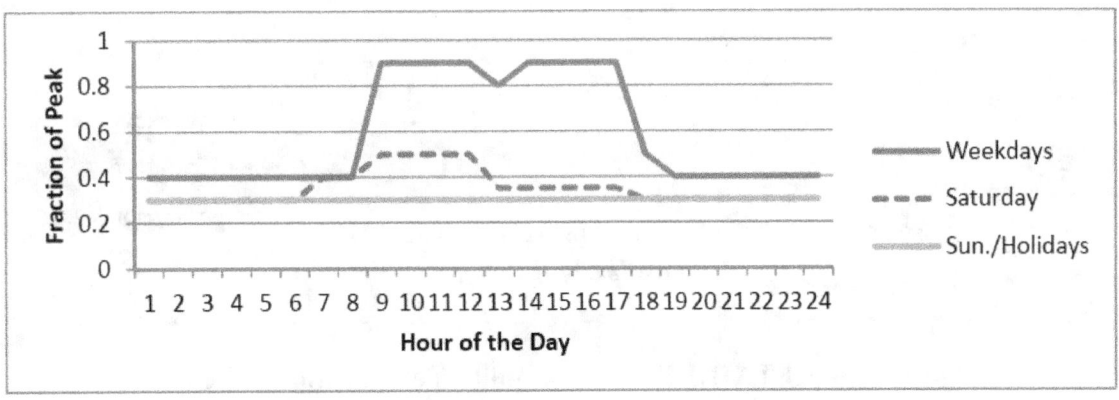

Figure 4-45 OFFIC16 Electrical Load Schedule

4.11 1-Story Retail Store

The 1-story retail store has mass walls, insulation entirely above the roof deck, fixed windows, and a window-to-wall ratio of 10 %. The detailed assumptions used in the energy simulations are described below.

4.11.1 Building Envelope

The energy efficiency characteristics of the building envelope are determined by the building's location and the edition of *ASHRAE 90.1*. The window characteristics (U-factor, SHGC, and VT) are based on the *ASHRAE 90.1* requirements for operable windows for 0.0 % to 10.0 % glazing. The wall and roof efficiency characteristics are based on the *ASHRAE 90.1* requirements for nonresidential buildings with above grade, mass wall construction and insulation entirely above the roof deck.

4.11.2 Heating, Ventilation, and Air Conditioning

There are four main aspects to the heating, ventilation, and air conditioning of a building: equipment, operating conditions, air infiltration, and mechanical ventilation. The HVAC equipment is a rooftop packaged electric air conditioner and a natural gas-fired furnace. Figure 4-46 show that the heating setpoint temperature for the retail store varies slightly by day of the week. For all days, the setpoint is 21°C (70°F) while the store is open and 15.6°C (60°F) when the store is closed. The store is open from 7 AM to 10 PM on weekdays and 7 AM to 11 PM on Saturdays. The store hour on Sundays and holidays are 9 AM to 8 PM. These setpoints correlate with the building occupancy schedule. Figure 4-47 shows a nearly identical pattern for the cooling setpoint temperature. The setpoint is 24°C (75°F) for 7 AM to 9 PM on weekdays, 7 AM to 10 PM on Saturdays, and 9 AM to 7 PM on Sundays and holidays. The setpoint is 30°C (86°F) while the store is closed.

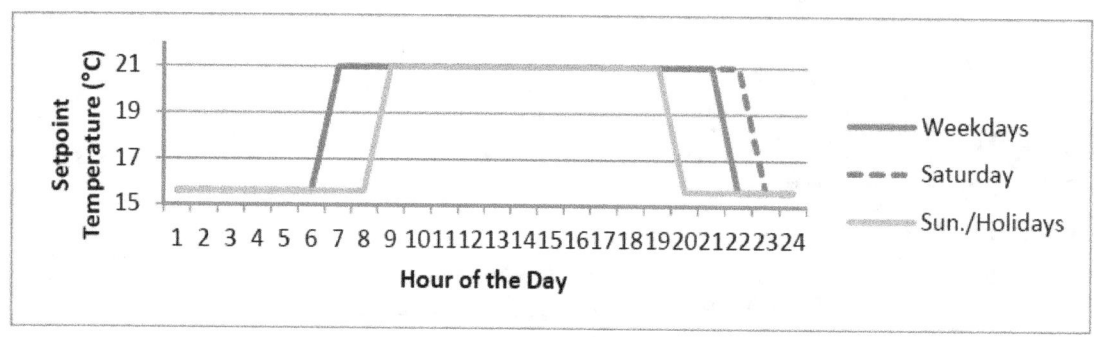

Figure 4-46 RETAIL1 Heating Setpoint Temperature Schedule

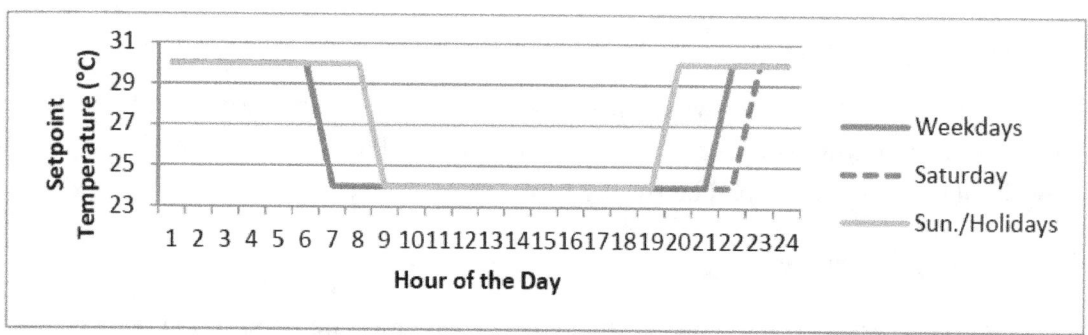

Figure 4-47 RETAIL1 Cooling Setpoint Temperature Schedule

Air infiltration and mechanical ventilation are assumed to be constant across all editions of *ASHRAE 90.1*, with an infiltration rate of 0.378 m^3/s per floor (0.43 ACH) and minimum mechanical ventilation of 0.547 m^3/s per floor (0.62 ACH).

4.11.3 Occupancy, Lighting, and Electrical Loads

The peak occupancy for the 1-story retail store is assumed to be 27 people or 1 person per 25.5 m^2 (300 ft^2). The schedule in Figure 4-48 shows that occupancy varies significantly both within a given day and across days of the week. In general, the afternoon and early evening is the busy time over all days, which is the most common time of the day for people to shop. The afternoon is the busiest time on the weekends and holidays while the early evening has the greatest occupancy on weekdays. The greatest occupancy occurs on Saturday followed by weekdays and then Sundays and holidays.

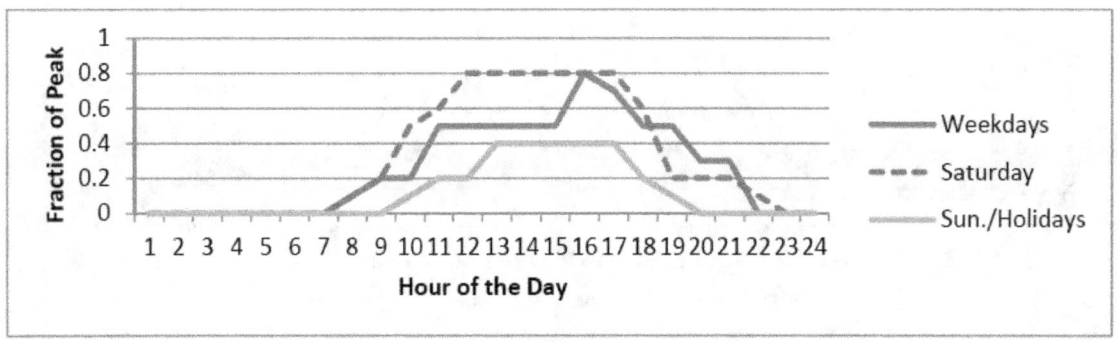

Figure 4-48 RETAIL1 Occupancy Schedule

The energy simulation assumes between 16.1 W/m^2 (1.5 W/ft^2) and 20.5 W/m^2 (1.9 W/ft^2) of lighting density depending on the building design (e.g. edition of *ASHRAE 90.1* or LEC). The lighting load schedules, as a fraction of peak lighting loads, in Figure 4-49 is highly correlated with the occupancy schedule. However, lighting is an on/off decision. So when the retail store is open, the lighting load is fairly constant and has less variability than the occupancy schedule.

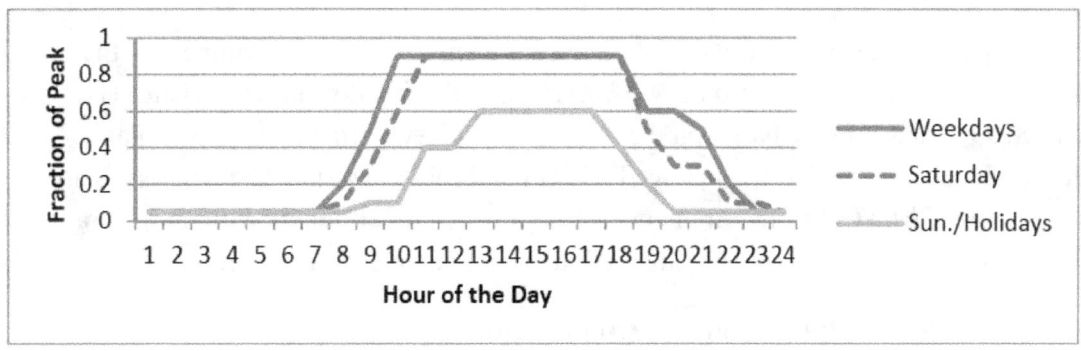

Figure 4-49 RETAIL1 Lighting Schedule

The peak electrical equipment load is 1999 W, or 2.69 W/m^2 (0.25 W/ft^2). The electrical load schedule in Figure 4-50 is highly correlated with the occupancy and lighting schedules, but has less variability across days of the week because most electrical loads are constant when a building is occupied.

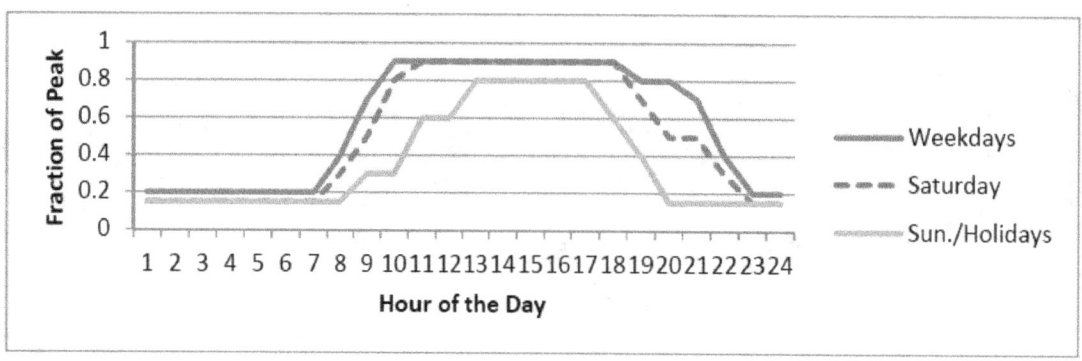

Figure 4-50 RETAIL1 Electrical Load Schedule

4.12 1-Story Restaurant

The 1-story restaurant has wood frame wall construction, insulation entirely above the roof deck, fixed windows, and a window-to-wall ratio of 30 %. The detailed assumptions used in the energy simulations are described below.

4.12.1 Building Envelope

The energy efficiency characteristics of the building envelope are determined by the building's location and the edition of *ASHRAE 90.1*. The window characteristics (U-factor, SHGC, and VT) are based on the *ASHRAE 90.1* requirements for operable windows for 20.1 % to 30.0 % glazing. The wall and roof efficiency characteristics are based on the *ASHRAE 90.1* requirements for nonresidential buildings with above grade, wood-framed wall construction and insulation entirely above the roof deck.

4.12.2 Heating, Ventilation, and Air Conditioning

There are four main aspects to the heating, ventilation, and air conditioning of a building: equipment, operating conditions, air infiltration, and mechanical ventilation. The HVAC equipment is a rooftop packaged electric air conditioner and a natural gas-fired furnace. Figure 4-51 show that the heating setpoint temperature for the retail store varies slightly by day of the week. The setpoint is 21°C (70°F) from 7 AM to 3 AM on weekdays, 9 AM to 3 AM on Saturdays, and 10 AM to 3 AM on Sundays and holidays. The heating setpoint is 15.6°C (60°F) for all other times. Figure 4-52 shows a mirror image for the cooling setpoint temperature. The setpoint is 24°C (75°F) from 7 AM to 3 AM on weekdays, 9 AM to 3 AM on Saturdays, and 10 AM to 3 AM on Sundays and holidays. The cooling setpoint is 30°C (86°F) for all other times.

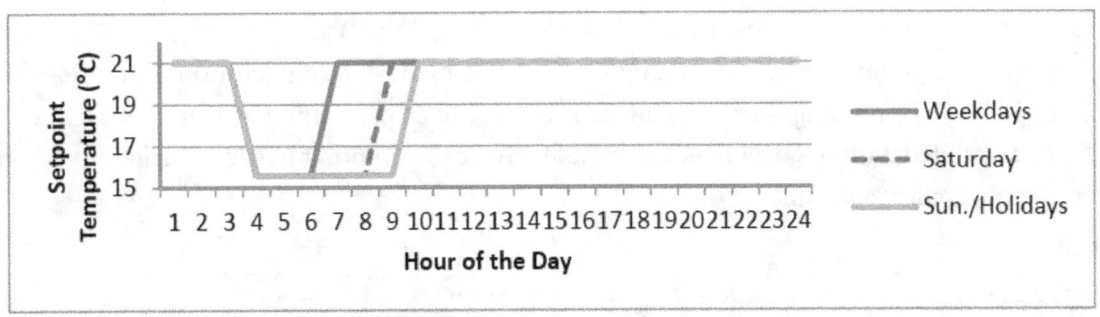

Figure 4-51 RSTRNT1 Heating Setpoint Temperature Schedule

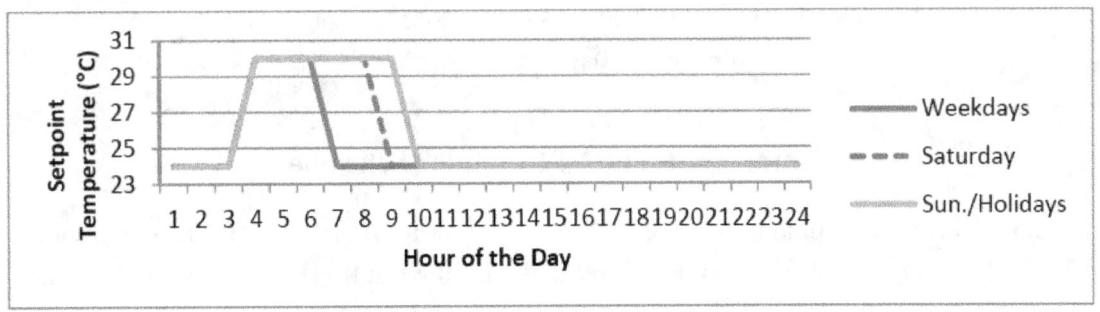

Figure 4-52 RSTRNT1 Cooling Setpoint Temperature Schedule

Air infiltration and mechanical ventilation are assumed to be constant across all editions of *ASHRAE 90.1*, with an infiltration rate of 0.275 m³/s per floor (0.58 ACH) and minimum mechanical ventilation of 0.609 m³/s per floor (1.29 ACH).

4.12.3 Occupancy, Lighting, and Electrical Loads

The peak occupancy for the 1-story restaurant is assumed to be 50 people or 1 person per 25.5 m² (100 ft²). The schedule in Figure 4-53 shows that occupancy varies significantly both within a given day and across days of the week. As would be expected, lunchtime and dinnertime are the busiest times over all days.

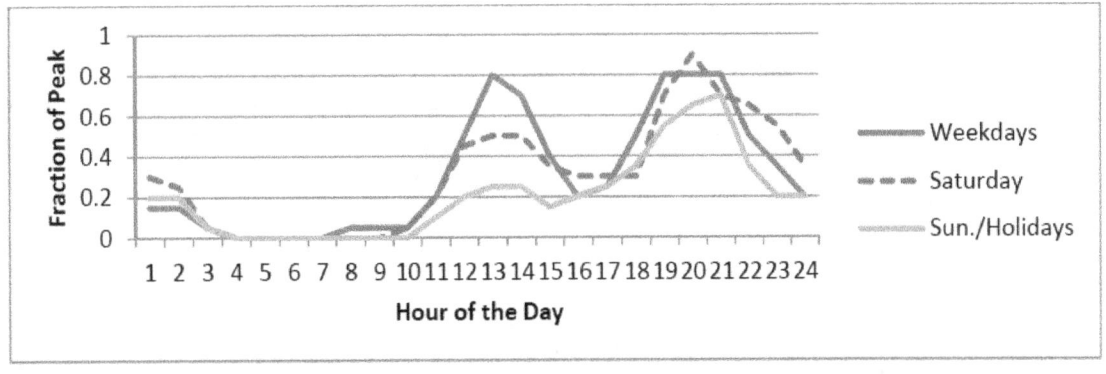

Figure 4-53 RSTRNT1 Occupancy Schedule

The energy simulation assumes between 14.0 W/m^2 (1.3 W/ft^2) and 19.4 W/m^2 (1.8 W/ft^2) of lighting density depending on the building design (e.g. edition of *ASHRAE 90.1* or LEC). The lighting load schedules, as a fraction of peak lighting loads, in Figure 4-54 is correlated with the occupancy schedule. However, lighting has less variability than the occupancy schedule.

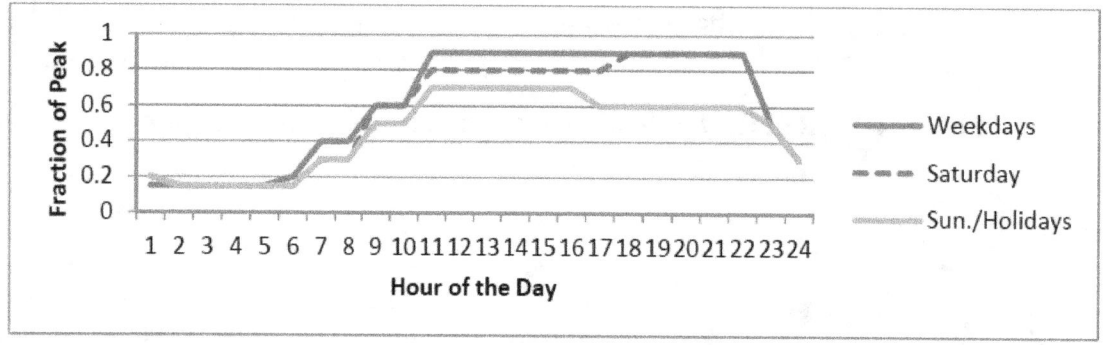

Figure 4-54 RSTRNT1 Lighting Schedule

The peak electrical equipment load is 502 W, or 1.08 W/m^2 (0.10 W/ft^2). The electrical load schedule in Figure 4-55 is highly correlated with the lighting schedules, but does not vary across days of the week because most electrical loads tend to be constant when a building is occupied no matter the amount of occupancy.

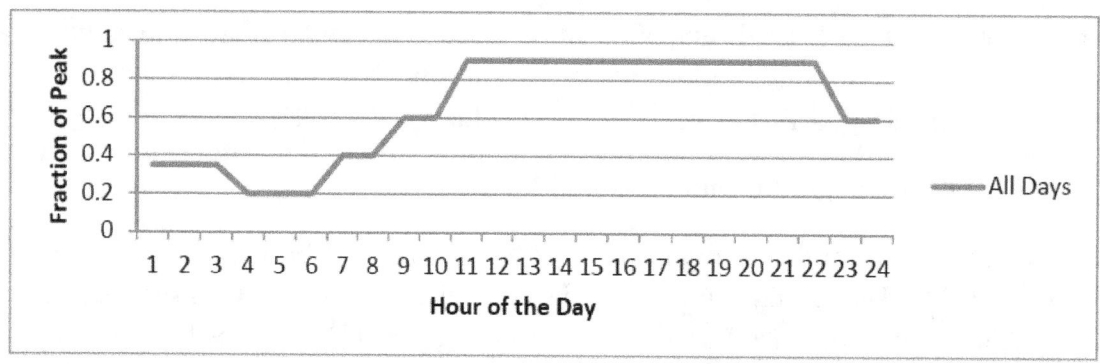

Figure 4-55 RSTRNT1 Electrical Load Schedule

5 Database Application and Expansion

This report defines the assumptions made to create the whole building energy simulations used to estimate the energy use for the new commercial building prototypes. The results of the whole building energy simulations documented in this report are combined with construction cost, maintenance, repair, and replacement cost, energy cost, and emissions data into a whole building sustainability database that allows the comparison of economic and environmental performance of different editions of *ASHRAE Standard 90.1*.

This documentation allows building energy researchers to better understand the underlying assumptions used to generate the energy simulation results built into the sustainability database. This report also allows future research based on the database to cite this documentation and focus on the analysis of results.

The initial 12 prototype commercial buildings are merely a starting point and can be expanded in a number of ways. First, additional types of commercial building prototypes can be added to increase the representation of the U.S. building stock, including the incorporation of the 16 NREL reference buildings. Second, additional sizes of the current commercial buildings prototypes can be included by adding floors or expanding the footprint of the current prototypes. Third, building components can be changed to better represent regional variation, such as adding electricity and fuel oil as heating fuel source options.

References

Field, Kristin, Deru, Michael, and Studer, Daniel. Using DOE Commercial Reference Buildings for Simulation Studies (Preprint). August 2010. NREL/CP-550-48588.

Department of Energy, Building Technologies Program, EnergyPlus energy simulation software, 2009, http://apps1.eere.energy.gov/buildings/energyplus/.

EnergyPlus Example File Generator, Building Energy Simulation Web Interface for EnergyPlus, accessed Feb. 2009, U.S. Department of Energy National Renewable Energy Laboratory, http://apps1.eere.energy.gov/buildings/energyplus/.

ASHRAE/IESNA Standard Project Committee 90.1, ASHRAE 90.1-1999 Standard- Energy Standard for Buildings Except Low-Rise Residential Buildings, 1999, ASHRAE, Inc.

ASHRAE/IESNA Standard Project Committee 90.1, ASHRAE 90.1-2001 Standard- Energy Standard for Buildings Except Low-Rise Residential Buildings, 2001, ASHRAE, Inc.

ASHRAE/IESNA Standard Project Committee 90.1, ASHRAE 90.1-2004 Standard- Energy Standard for Buildings Except Low-Rise Residential Buildings, 2004, ASHRAE, Inc.

ASHRAE/IESNA Standard Project Committee 90.1, ASHRAE 90.1-2007 Standard- Energy Standard for Buildings Except Low-Rise Residential Buildings, 2007, ASHRAE, Inc.

Database of State Incentives for Renewables and Efficiency, Rules, Regulations, and Policies for Energy Efficiency database, building energy codes, accessed summer 2010, www.dsireusa.org/.

Department of Energy, Building Technologies Program, State energy codes at-a-glance, 2009, http://www.energycodes.gov/states/maps/commercialStatus.stm.

Commercial Buildings Energy Consumption Survey database, 2003, accessed June-July 2009, http://www.eia.doe.gov/emeu/cbecs/.

RS Means CostWorks databases, accessed Oct. 2008-March 2009, http://www.meanscostworks.com/.

Appendix A   Selected File Input and Output Formats

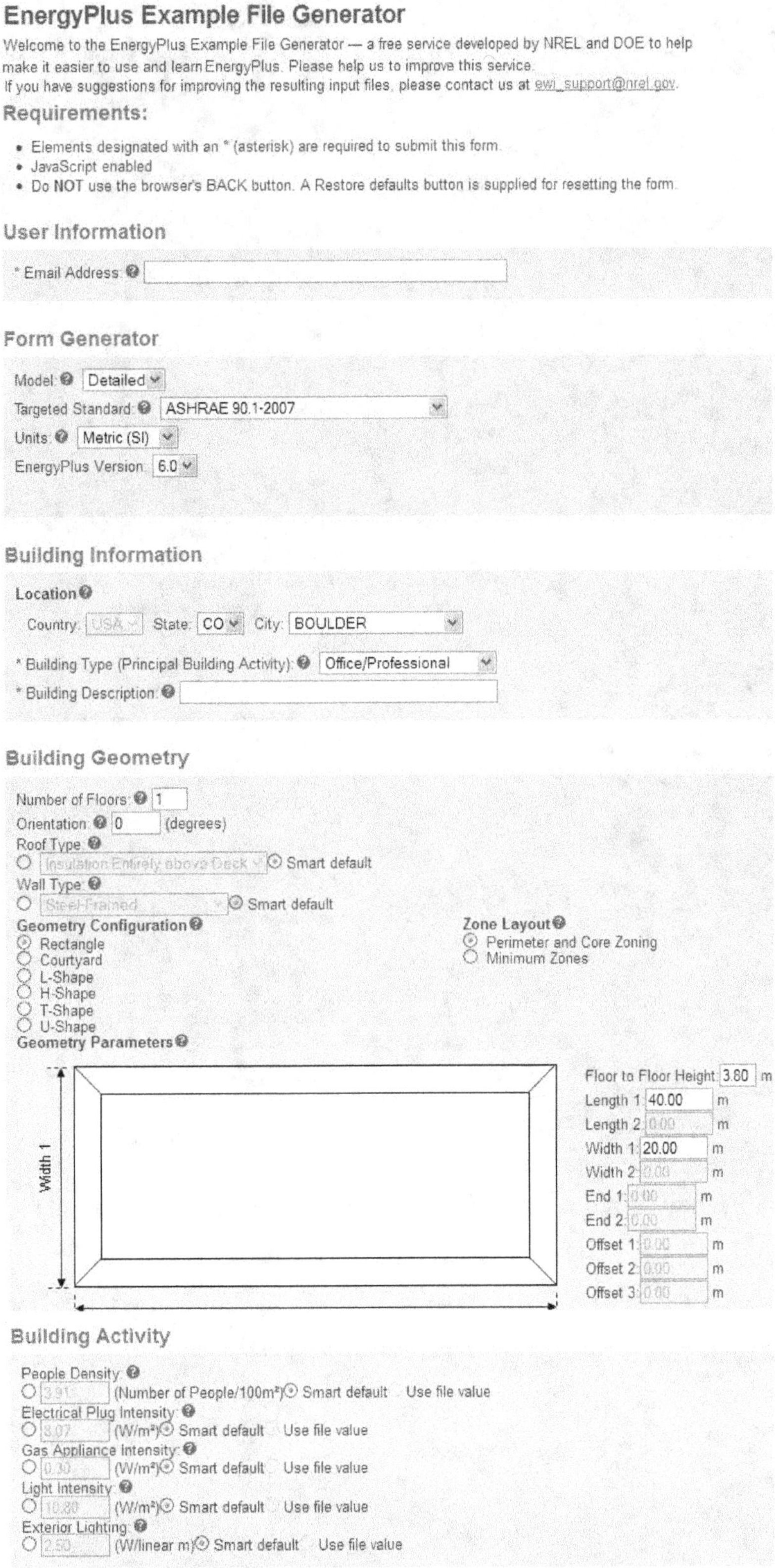

EnergyPlus Example File Generator

Welcome to the EnergyPlus Example File Generator — a free service developed by NREL and DOE to help make it easier to use and learn EnergyPlus. Please help us to improve this service.
If you have suggestions for improving the resulting input files, please contact us at ewi_support@nrel.gov.

Requirements:

- Elements designated with an * (asterisk) are required to submit this form.
- JavaScript enabled
- Do NOT use the browser's BACK button. A Restore defaults button is supplied for resetting the form.

User Information

* Email Address:

Form Generator

Model: Detailed
Targeted Standard: ASHRAE 90.1-2007
Units: Metric (SI)
EnergyPlus Version: 6.0

Building Information

Location
   Country: USA   State: CO   City: BOULDER

* Building Type (Principal Building Activity): Office/Professional
* Building Description:

Building Geometry

Number of Floors: 1
Orientation: 0 (degrees)
Roof Type:
  ○ Insulation Entirely above Deck   ⊙ Smart default
Wall Type:
  ○ Steel-Framed   ⊙ Smart default

Geometry Configuration
  ⊙ Rectangle
  ○ Courtyard
  ○ L-Shape
  ○ H-Shape
  ○ T-Shape
  ○ U-Shape

Zone Layout
  ⊙ Perimeter and Core Zoning
  ○ Minimum Zones

Geometry Parameters

Floor to Floor Height:	3.60 m
Length 1:	40.00 m
Length 2:	0.00 m
Width 1:	20.00 m
Width 2:	0.00 m
End 1:	0.00 m
End 2:	0.00 m
Offset 1:	0.00 m
Offset 2:	0.00 m
Offset 3:	0.00 m

Building Activity

People Density:
  ○ 3.91 (Number of People/100m²) ⊙ Smart default   Use file value
Electrical Plug Intensity:
  ○ 8.07 (W/m²) ⊙ Smart default   Use file value
Gas Appliance Intensity:
  ○ 0.30 (W/m²) ⊙ Smart default   Use file value
Light Intensity:
  ○ 10.80 (W/m²) ⊙ Smart default   Use file value
Exterior Lighting:
  ○ 2.50 (W/linear m) ⊙ Smart default   Use file value

Figure A-1 Example File Generator Inputs

Building Fenestration

○ Use Custom Values for all Building Fenestration inputs ◉ Smart default Use file value

South

Glazing Percentage: ❷ [40] (%)
Overhang ❷
Projection Factor: [0.00] Offset: [0.0] (m)
Fin ❷
Projection Factor: [0.00] Offset: [0.0] (m)

West

Glazing Percentage: ❷ [40] (%)
Overhang ❷
Projection Factor: [0.00] Offset: [0.0] (m)
Fin ❷
Projection Factor: [0.00] Offset: [0.0] (m)

North

Glazing Percentage: ❷ [40] (%)
Overhang ❷
Projection Factor: [0.00] Offset: [0.0] (m)
Fin ❷
Projection Factor: [0.00] Offset: [0.0] (m)

East

Glazing Percentage: ❷ [40] (%)
Overhang ❷
Projection Factor: [0.00] Offset: [0.0] (m)
Fin ❷
Projection Factor: [0.00] Offset: [0.0] (m)

Roof Fenestration (choose one)

◉ None
○ Skylights
　[ ] Percentage of roof area (%)
○ Tubular Daylighting Devices
　[18.00] TDD density (m²/TDD)

Daylighting

Daylighting Illuminance Setpoint: ❷ [398] (lux)
Control Type: ❷ [Stepped ▾] Number of Steps: [3]

Building HVAC System

Select System based on: ❷
◉ 90.1-2004 Appendix G Types
○ Heating/Cooling Type
ASHRAE 90.1-2004 Appendix G Types: [Automatically select based on building ▾]
Heating Type: [Furnace ▾]
Cooling Type: [Evaporative Cooler ▾]

Fan Static Pressure: ❷
○ [2.01] (Pa) ◉ Smart default
Total Fan Efficiency: ❷
○ [60] (%) ◉ Smart default
Cooling COP: ❷
　[3.0] ◉ Smart default
Heating Efficiency: ❷
　[80] (%) ◉ Smart default
Heating COP: ❷
　[3.0] ◉ Smart default
Zone/Terminal Fan Static Pressure: ❷
○ [301.18] (Pa) ◉ Smart default

Outside Air

Outside Air Ventilation Per Person: ❷
○ [2.5] (L/sec) ◉ Smart default
Outside Air Ventilation Per Area: ❷
○ [0.30] (L/sec/m²) ◉ Smart default
Annual Avg. Infiltration Rate: ❷
○ [0.5] (1/hr) ◉ Smart default

Service Water Heating

Load

Water Use: ❷ [0.0000] (Liters) [per hour ▾] [40] (°C)

Supply

Storage Set Point: ❷ [60] (°C)
[Natural Gas ▾] [80] (Thermal Efficiency %)

Photovoltaics

Installation Type ❷
◉ None
○ Area Percentage
[5] (%)
○ Installed Capacity
[0.00] (W)

Cell Efficiency: ❷ [10.0] (%)
Inverter Efficiency: ❷ [89] (%)
Coverage Surfaces: ❷ [Roof Only ▾]

Figure A-1 Example File Generator Inputs (continued)

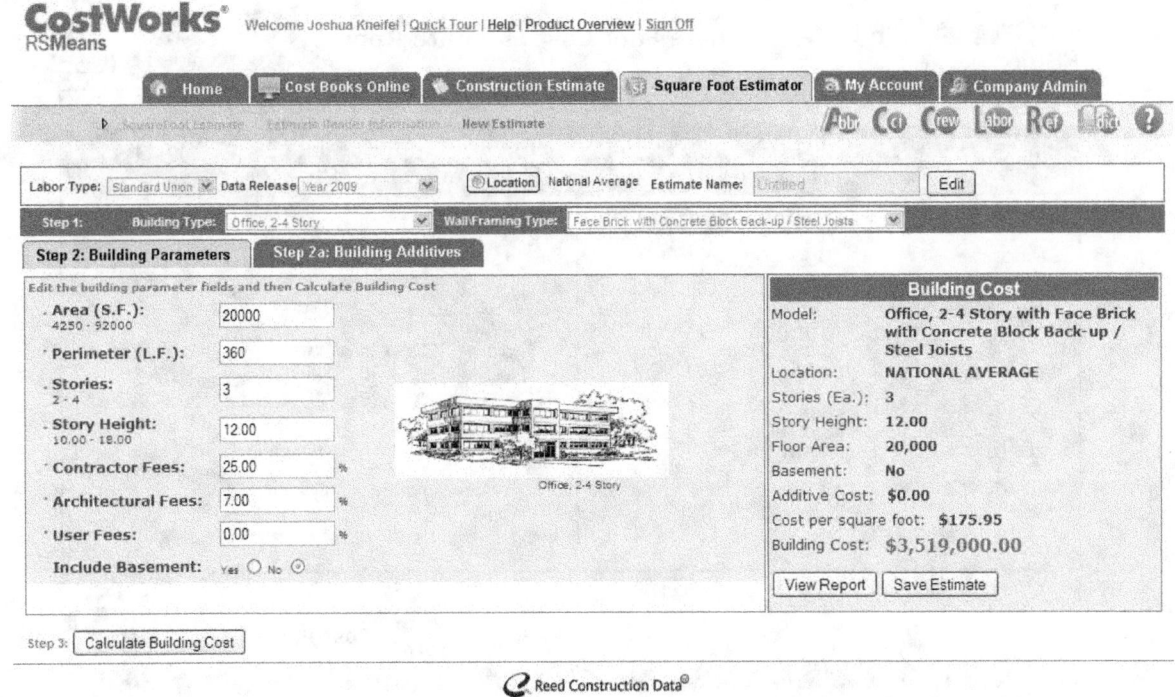

Figure A-2 RSMeans CostWorks Square Foot Cost Estimator Interface Input

Estimate Name: **Untitled**	
Building Type:	**Office, 2-4 Story with Face Brick with Concrete Block Back-up / Steel Joists**

Location:	**National Average**
Stories:	**3**
Story Height (L.F.):	**12.00**
Floor Area (S.F.):	**20000**
Labor Type:	**Union**
Basement Included:	**No**
Data Release:	**Year 2009**
Cost Per Square Foot:	
Building Cost:	

Costs are derived from a building model with basic components. Scope differences and market conditions can cause costs to vary significantly.

		% of Total	Cost Per S.F.	Cost
A Substructure		**4.4%**		
A1010	**Standard Foundations**			
	Strip footing, concrete, reinforced, load 11.1 KLF, soil bearing capacity 6 KSF, 12" deep x 24" wide			
	Spread footings, 3000 PSI concrete, load 200K, soil bearing capacity 6 KSF, 6'- 0" square x 20" deep			
	Spread footings, 3000 PSI concrete, load 300K, soil bearing capacity 6 KSF, 7'- 6" square x 25" deep			
A1030	**Slab on Grade**			
	Slab on grade, 4" thick, non industrial, reinforced			
A2010	**Basement Excavation**			
	Excavate and fill, 30,000 SF, 4' deep, sand, gravel, or common earth, on site storage			
A2020	**Basement Walls**			
	Foundation wall, CIP, 4' wall height, direct chute, .099 CY/LF, 4.8 PLF, 8" thick			
	Foundation wall, CIP, 4' wall height, direct chute, .148 CY/LF, 7.2 PLF, 12" thick			
B Shell		**29.6%**		
B1010	**Floor Construction**			
	Floor, concrete, slab form, open web bar joist @ 2' OC, on W beam and wall, 25'x25' bay, 26" deep, 75 PSF superimposed load, 120 PSF total load			
	Floor, concrete, slab form, open web bar joist @ 2' OC, on W beam and wall, 25'x25' bay, 26" deep, 75 PSF superimposed load, 120 PSF total load, for columns add			
	Fireproofing, gypsum board, fire rated, 2 layer, 1" thick, 14" steel column, 3 hour rating, 22 PLF			
B1020	**Roof Construction**			
	Floor, steel joists, beams, 1.5" 22 ga metal deck, on columns and bearing wall, 25'x25' bay, 20" deep, 40 PSF superimposed load, 60 PSF total load			
	Floor, steel joists, beams, 1.5" 22 ga metal deck, on columns and bearing wall, 25'x25' bay, 20" deep, 40 PSF superimposed load, 60 PSF total load, add for column			
B2010	**Exterior Walls**			
	Brick wall, composite double wythe, standard face/CMU back-up, 8" thick, perlite core fill			
B2020	**Exterior Windows**			
	Windows, aluminum, awning, insulated glass, 4'-5" x 5'-3"			

Figure A-3 RSMeans CostWorks Square Foot Cost Estimator Interface Output

Appendix B   Selected Tabular Information

Table B-1 Cities Included in the Simulation Database

State	City	State	City	State	City	State	City
AK	Anchorage	IA	Waterloo	MT	Missoula	PA	Wilkes-Barre
AK	Barrow	ID	Boise	NC	Asheville	PA	Williamsport
AK	Fairbanks	ID	Lewiston	NC	Charlotte	RI	Providence
AK	Juneau	ID	Pocatello	NC	Greensboro	SC	Charleston
AK	Kodiak	IL	Chicago	NC	Hatteras	SC	Columbia
AK	Nome	IL	Moline	NC	Raleigh	SC	Greenville
AL	Birmingham	IL	Peoria	NC	Wilmington	SD	Huron
AL	Huntsville	IL	Rockford	ND	Bismarck	SD	Pierre
AL	Mobile	IL	Springfield	ND	Fargo	SD	Sioux Falls
AL	Montgomery	IN	Evansville	ND	Minot	TN	Bristol
AR	Fort Smith	IN	Fort Wayne	NE	Grand Island	TN	Chattanooga
AR	Little Rock	IN	Indianapolis	NE	Norfolk	TN	Knoxville
AZ	Flagstaff	IN	South Bend	NE	North Platte	TN	Memphis
AZ	Phoenix	KS	Dodge City	NE	Omaha	TN	Nashville
AZ	Prescott	KS	Goodland	NE	Scottsbluff	TX	Abilene
AZ	Tucson	KS	Topeka	NH	Concord	TX	Amarillo
AZ	Winslow	KS	Wichita	NJ	Atlantic City	TX	Austin
AZ	Yuma	KY	Covington	NJ	Newark	TX	Brownsville
CA	Arcata	KY	Lexington-Fayette	NM	Albuquerque	TX	Corpus Christi
CA	Bakersfield	KY	Louisville	NM	Roswell	TX	Del Rio
CA	Daggett	LA	Baton Rouge	NM	Tucumcari	TX	El Paso
CA	Fresno	LA	Lake Charles	NV	Elko	TX	Fort Worth
CA	Long Beach	LA	New Orleans	NV	Ely	TX	Houston
CA	Los Angeles	LA	Shreveport	NV	Las Vegas	TX	Lubbock
CA	Riverside	MA	Boston	NV	Reno	TX	Lufkin
CA	Sacramento	MA	Worcester	NV	Tonopah	TX	Midland
CA	San Diego	MD	Baltimore	NV	Winnemucca	TX	Port Arthur
CA	San Francisco	ME	Caribou	NY	Albany	TX	San Angelo
CA	Santa Maria	ME	Portland	NY	Binghamton	TX	San Antonio
CO	Alamosa	MI	Alpena	NY	Buffalo	TX	Victoria
CO	Boulder	MI	Detroit	NY	Massena	TX	Waco
CO	Colorado Springs	MI	Flint	NY	New York	TX	Wichita Falls
CO	Eagle	MI	Grand Rapids	NY	Rochester	UT	Cedar City
CO	Grand Junction	MI	Houghton	NY	Syracuse	UT	Salt Lake City
CO	Pueblo	MI	Lansing	OH	Akron	VA	Lynchburg
CT	Bridgeport	MI	Muskegon	OH	Cleveland	VA	Norfolk
CT	Hartford	MI	Sault Sainte Marie	OH	Columbus	VA	Richmond
DE	Wilmington	MI	Traverse City	OH	Mansfield	VA	Roanoke
FL	Daytona Beach	MN	Duluth	OH	Toledo	VT	Burlington
FL	Jacksonville	MN	International Falls	OH	Youngstown	WA	Olympia
FL	Key West	MN	Minneapolis	OK	Oklahoma City	WA	Quillayute
FL	Miami	MN	Rochester	OK	Tulsa	WA	Seattle
FL	Tallahassee	MN	Saint Cloud	OR	Astoria	WA	Spokane
FL	Tampa	MO	Columbia	OR	Burns	WA	Yakima
FL	West Palm Beach	MO	Kansas City	OR	Eugene	WI	Eau Claire
GA	Athens	MO	Saint Louis	OR	Medford	WI	Green Bay
GA	Atlanta	MO	Springfield	OR	North Bend	WI	La Crosse
GA	Augusta	MS	Jackson	OR	Pendleton	WI	Madison
GA	Columbus	MS	Meridian	OR	Portland	WI	Milwaukee
GA	Macon	MT	Billings	OR	Redmond	WV	Charleston
GA	Savannah	MT	Cut Bank	OR	Salem	WV	Elkins

Table B-1 Cities Included in the Simulation Database (continued)

State	City	State	City	State	City	State	City
HI	Hilo	MT	Glasgow	PA	Allentown	WV	Huntington
HI	Honolulu	MT	Great Falls	PA	Bradford	WY	Casper
IA	Burlington	MT	Helena	PA	Erie	WY	Cheyenne
IA	Des Moines	MT	Kalispell	PA	Harrisburg	WY	Lander
IA	Mason City	MT	Lewistown	PA	Philadelphia	WY	Rock Springs
IA	Sioux City	MT	Miles City	PA	Pittsburgh	WY	Sheridan

Table B-2 CBECS Categories and Subcategories

Category	Subcategory	Category	Subcategory
Education	elementary or middle school high school college or university preschool or daycare adult education career or vocational training religious education	Public Assembly	social or meeting recreation entertainment or culture library funeral home student activities center armory exhibition hall broadcasting studio transportation terminal
Food Sales	grocery store or food market gas station with a convenience store; convenience store		
Food Service	fast food restaurant or cafeteria	Public Order and Safety	police station fire station jail, reformatory, or penitentiary courthouse or probation office
Health Care Inpatient	hospital inpatient rehabilitation		
		Religious Worship	None
Health Care Outpatient	medical office (see previous column) clinic or other outpatient health care outpatient rehabilitation veterinarian	Service	vehicle service or vehicle repair shop vehicle storage/ maintenance (car barn) repair shop dry cleaner or laundromat post office or postal center car wash gas station photo processing shop beauty parlor or barber shop tanning salon copy center or printing shop kennel
Lodging	motel or inn hotel dormitory, fraternity, or sorority retirement home nursing home, assisted living, etc. convent or monastery shelter, orphanage, halfway house		
Mercantile Non-Mall	retail store beer, wine, or liquor store rental center dealership or showroom for vehicles or boats studio/gallery	Warehouse and Storage	refrigerated warehouse non-refrigerated warehouse distribution or shipping center
Mercantile Malls	enclosed mall strip shopping center	Other	airplane hangar crematorium laboratory telephone switching agricultural with some retail space manufacturing or industrial with some retail space data center or server farm
Office	administrative or professional office government office mixed-use office bank or other financial institution medical office (see previous column) sales office contractor's office non-profit or social services research and development city hall or city center religious office call center	Vacant	None